Selected Papers from the 9th Greek Conference of Biochemistry and Physiology of Exercise

Selected Papers from the 9th Greek Conference of Biochemistry and Physiology of Exercise

Editor

Vassilis Mougios

MDPI • Basel • Beijing • Wuhan • Barcelona • Belgrade • Manchester • Tokyo • Cluj • Tianjin

Editor
Vassilis Mougios
Aristotle University of Thessaloniki
Greece

Editorial Office
MDPI
St. Alban-Anlage 66
4052 Basel, Switzerland

This is a reprint of articles from the Special Issue published online in the open access journal *Sports* (ISSN 2075-4663) (available at: https://www.mdpi.com/journal/sports/special_issues/9GCBPE).

For citation purposes, cite each article independently as indicated on the article page online and as indicated below:

LastName, A.A.; LastName, B.B.; LastName, C.C. Article Title. *Journal Name* **Year**, *Volume Number*, Page Range.

ISBN 978-3-0365-0428-5 (Hbk)
ISBN 978-3-0365-0429-2 (PDF)

Contents

About the Editor

Vassilis Mougios is Professor of Exercise Biochemistry and Director of the Laboratory of Evaluation of Human Biological Performance at the School of Physical Education and Sport Science, Aristotle University of Thessaloniki, Greece. He has been teaching exercise biochemistry, exercise physiology, sport nutrition, and ergogenic aspects of sport to undergraduate and graduate audiences in his home country and abroad for over 30 years. Mougios has been invited to numerous conferences and conducted extensive research on muscle activity, exercise metabolism, biochemical assessment of athletes, and sport nutrition. As a result, he has authored or coauthored over 100 papers in international scientific journals. He is the author of *Exercise Biochemistry*, a book published by Human Kinetics and presently in its 2nd edition. Mougios is a fellow and member of the reviewing panel of the European College of Sport Science. He has served or serves as associate editor for 3 journals, topic editor for another 5 journals, and reviewer for over 40 journals.

Preface to "Selected Papers from the 9th Greek Conference of Biochemistry and Physiology of Exercise"

Exercise biochemistry and physiology is a very active field of scientific endeavor, which, over the past half-century, has contributed immensely to our understanding of how muscle activity alters the way our bodies (and those of other animals) function. On the occasion of its 9th annual conference (held in Thessaloniki, Greece, in October 2019), the Hellenic Society of Biochemistry and Physiology of Exercise invited a number of presenters to publish their work as a full paper in a Special Issue in *Sports*. The present volume is the fruit of that endeavor. It contains 10 original papers on biochemical and physiological adaptations to training in children, adolescents, and adults; exercise testing; and exercise as medicine.

We hope that the present collection will satisfy interested readers and will stimulate fruitful discussion and ideas for future research.

Vassilis Mougios
Editor

Article

Heart Rate Responses during Sport-Specific High-Intensity Circuit Exercise in Child Female Gymnasts

Andreas Salagas, Olyvia Donti, Christos Katsikas and Gregory C. Bogdanis *

School of Physical Education and Sports Science, National and Kapodistrian University of Athens, 172 37 Athens, Greece; andreassal@phed.uoa.gr (A.S.); odonti@phed.uoa.gr (O.D.); ckatsikas@phed.uoa.gr (C.K.)

* Correspondence: gbogdanis@phed.uoa.gr; Tel.: +30-210-727-6115

Received: 9 April 2020; Accepted: 15 May 2020; Published: 18 May 2020

Abstract: This study examined heart rate (HR) responses during a sport-specific high-intensity circuit training session to indirectly assess cardiorespiratory stress in child athletes. Seventeen, female gymnasts, aged 9–11 years performed two 5-min 15 s sets of circuit exercise, interspersed by a 3 min rest interval. Each set included five rounds of five gymnastic exercises (7 s work, 7 s rest) executed with maximal effort. During the first circuit training set, peak heart rate (HR) was 192 ± 7 bpm and average HR was 83 ± 4% of maximum HR (HR_{max}), which was determined in a separate session. In the second set, peak HR and average HR were increased to 196 ± 8 bpm ($p < 0.001$, $d = 0.55$) and to 89 ± 4% HR_{max} ($p < 0.001$, $d = 2.19$), respectively, compared with the first set. HR was above 80% HR_{max} for 4.1 ± 1.2 min during set 1 and this was increased to 5.1 ± 0.4 min in set 2 ($p < 0.001$, $d = 1.15$). Likewise, HR was above 90% of HR_{max} for 2.0 ± 1.2 min in set 1 and was increased to 3.4 ± 1.7 min in set 2 ($p < 0.001$, $d = 0.98$). In summary, two 5-min 15 s sets of high-intensity circuit training using sport-specific exercises, increased HR to levels above 80% and 90% HR_{max} for extended time periods, and thus may be considered as an appropriate stimulus, in terms of intensity, for improving aerobic fitness in child female gymnasts.

Keywords: aerobic fitness; intermittent exercise; prepubertal children

1. Introduction

A large number of studies over the last decade have shown that high-intensity interval training (HIIT) improves athletic performance and health in adults [1–3]. HIIT typically includes short duration exercise bouts (15–60 s) performed at an intensity around maximal oxygen uptake (VO_{2max}) [4,5], or shorter bouts (6–15 s) executed at intensities corresponding to 100–130% of VO_{2max} with work-to-rest ratios of 1:1 to 1:1.5 [4,6]. Previous studies have shown significant aerobic contribution during high-intensity exercise in adults [7], while children demonstrate even higher reliance on their aerobic metabolism in this type of exercise, due to their faster VO_2 kinetics and lower glycolytic energy supply [8,9]. However, the majority of HIIT studies in children used intense running or cycling [5,8,10,11], and little is known regarding the aerobic contribution during other forms of high-intensity exercise, such as functional training, with most data obtained from adult populations [12,13]. This type of training is commonly used by coaches in many sports, such as gymnastics, and typically includes sport-specific exercises using body weight, which are executed in a circuit fashion, aiming to improve neuromuscular performance [14]. However, there is very limited information about the acute cardiorespiratory stress for this type of circuit training program, especially in child athletes who are systematically training from a very young age [15].

Artistic gymnastics is a popular sport that requires high levels of strength, power, flexibility, coordination and anaerobic power [14]. During training sessions, gymnastic routines and exercises are performed repetitively over a long period of time, with short recovery intervals. Thus, the ability to recover is important not only to preserve a high quality of technical execution throughout a training session, but also for optimal performance during competitions [14]. Previous studies have shown that aerobic fitness is an important determinant of performance recovery, while the aerobic contribution to energy supply is substantial when high-intensity efforts last 20–30 s or longer [7,16]. For example, in artistic gymnastics, VO_2 during competitive routines of floor exercises lasting 90 s, is increased to 85% of maximal VO_2 (VO_{2max}) [17]. Similarly, peak heart rate reaches values over 90% of maximum heart rate in all apparatuses except the vault, where heart rate does not increase to such high levels, due to the short duration of a single vault (5–7 s) [17]. Furthermore, VO_{2max} explained 92.5% of the variation in performance scores in elite rhythmic gymnasts, whose competitive routines last about 60–90 s [18]. Thus, aerobic fitness may be important for artistic gymnasts' performance, and although high-intensity circuit training using sport-specific exercises is used by gymnastics coaches from an early age [14,19], evidence is limited regarding the physiological stress imposed on the cardio-respiratory system of developing athletes. Thus, the aim of this study was to examine heart rate responses during a high-intensity circuit training session using sport-specific exercises in child female gymnasts.

2. Materials and Methods

2.1. Participants

Participants were recruited from a local gymnastics club. The inclusion criteria were: (1) healthy female athletes, (2) participation in competitive artistic gymnastics for 3–4 years, (3) weekly training for at least 4 hours, and (4) age range between 9–11 years. Athletes who had any musculoskeletal injury from the previous 6 months were excluded from the study. Seventeen premenarcheal female artistic gymnasts aged 9.7 ± 0.8 years, with body mass 33.7 ± 7.3 kg, height 1.38 ± 0.10 cm and Body Mass Index 17.4 ± 2.4 kg/m^2, participated in the testing procedures. All procedures were in accordance with the Declaration of Helsinki and approved by the local university ethics committee (approval no. 1198). Parents and participants were informed about the experimental procedure and signed an informed consent. All participants had an athlete's health card validated by the Hellenic Gymnastics Federation.

2.2. Procedure

The experimental protocol was performed in the pre-season. Following two familiarization sessions performed 2–3 days apart, the participants executed a 20 m shuttle run test until exhaustion, to measure maximal heart rate (HR_{max}) and to estimate maximal oxygen consumption (VO_{2max}), using a standardized age-specific equation [20]. Three days after the shuttle run test, the main testing protocol was performed, which consisted of two sets of five gymnastic exercises executed in a circuit manner (Table 1). Participants abstained from any rigorous physical activity for 24 h before testing Each exercise was performed for 7 s, followed by a rest interval of equal duration, during which athletes moved to the next exercise (Table 1). These five exercises were executed in a circuit fashion until a total of 5 rounds was completed. Thus, each set included 5 rounds of 5 exercises with a total duration of 5 min and 15 s. A passive recovery period of 3 min separated the two sets. A 5 min standardized, sport-specific warm-up preceded the circuit training session, followed by 3 min of rest. The warm-up included 3 min of light jogging and 2 min of mobility exercises.

Table 1. High-intensity circuit training program executed 5 times.

Exercise Description	Number of Repetitions/Duration
1. From hanging on high bar leg raise in tuck position to dislocate in eagle grip (L-grip), release and land on the floor	3 reps/7 s
2. From cross support facing the end of a low beam (20 cm), lateral jumps across the length of the beam	4 reps/7 s
3. Forward roll, jump with half turn (180°), backward roll and jump with half turn (180°) (without pause or extra steps)	3 reps/7 s
4. From front support on parallel bars, forward swing to straddle position and straddle travel across the length of the parallel bars	3 reps/7 s
5. From front support on low bar cast backward to horizontal	3 reps/7 s

2.3. Heart Rate Measurements

During the 20 m shuttle run test, subjects' heart rate was monitored continuously using online telemetry (Polar Team 2, Polar Electro Oy, Kempele, Finland). During the main testing procedure, heart rate (HR) was being measured continuously (every 1 s) for the entire duration of the protocol, i.e., during set 1 and set 2, including 3 min of recovery after each set (see Figure 1). From the heart rate data, the following parameters were extracted or calculated: (a) peak HR, (b) mean HR, (c) time during which heart rate was above 80% of HR_{max}, (d) time during which heart rate was above 90% of HR_{max}, (e) heart rate recovery 1 and 2 min after each set of the circuit training (i.e., the drop of HR at the respective time points compared with the peak attained in each set) [21,22].

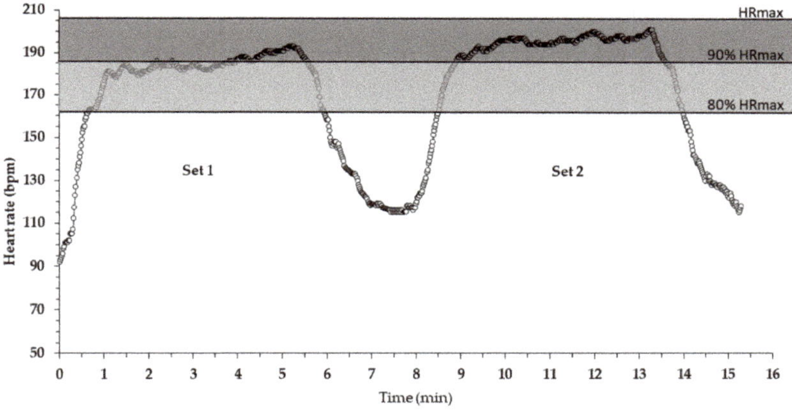

Figure 1. Time course of heart rate during the two sets of circuit training, separated by a 3 min rest interval, for one of the participants.

2.4. Statistical Analysis

Data analysis was performed using SPSS Statistics (Ver. 25, IBM Corporation, New York, NY, USA). Descriptive statistics were calculated (mean values and standard deviations). Comparisons between the heart rate variables of the first and the second test were performed using a paired-sample T-test. Effect sizes were determined by Cohen's d (trivial: 0–0.19, small: 0.20–0.49, medium: 0.50–0.79 and large: 0.80 and greater) [23]. One-way analysis of variance (ANOVA) followed by Tukey's post-hoc test, was used to examine whether peak heart rate and heart rate recovery observed during the two sets of circuit exercise session were different from the respective values (i.e., maximal heart rate and heart rate recovery) recorded during the shuttle run test. Significance was accepted at $p < 0.05$.

3. Results

3.1. Circuit Exercise Session

Peak heart rate reached in set 1 of the circuit exercise training session was 92 ± 4% of HR_{max}, and was increased to 95 ± 4% of HR_{max} in set 2 (Table 2). Mean heart rate during the first set of the circuit exercise session was 83 ± 4% HR_{max} and was further increased to 89 ± 4% HR_{max} in set 2 (Table 2). Moreover, the time during which HR was above 80% HR_{max} and 90% HR_{max} was higher in set 2 compared with set 1 ($p < 0.00$, $d = 1.15$) ($p < 0.001$, $d = 0.98$) (Table 2). The time course of heart rate during the high-intensity circuit training session from a representative individual is shown in Figure 1.

Table 2. Comparison of heart rate responses during the first and the second set (set 1 and set 2) of the high-intensity circuit exercise training session.

Header	Set 1	Set 2	*p* Value	Cohen's d
Peak heart rate (bpm)	192 ± 7	196 ± 8	<0.001	0.55
Mean heart rate (bpm)	171 ± 8	186 ± 6	<0.001	2.19
Time spent >80% [1] HR_{max} (min)	4.11 ± 1.19	5.09 ± 0.36	<0.001	1.15
Time spent >90% [1] HR_{max} (min)	2.01 ± 1.16	3.36 ± 1.65	<0.001	0.98
1 min heart rate recovery (bpm)	54 ± 13	54 ± 12	0.918	0.00
2 min heart rate recovery (bpm)	72 ± 13	69 ± 13	0.273	0.24

[1] HR_{max}: maximum heart rate attained during the shuttle run test.

3.2. Shuttle Run Test

HR_{max} attained during the shuttle-run test was 207 ± 5 beats per minute (bpm), while the estimated VO_{2max} was 49 ± 3 mL/kg/min HR_{max} attained during the shuttle run test to exhaustion ($p < 0.001$) was 5–8% higher compared with the peak HR during set 1 and set 2 of the circuit training exercise session. However, the recovery of HR after the shuttle run test was similar to HR recovery observed in set 1 and set 2 of the circuit exercise training session (1st min: 58 ± 16 and 2nd min: 76 ± 13 bpm, $p = 0.26$ to 0.65, Table 2).

4. Discussion

The main finding of this study was that this sport-specific high-intensity circuit training, which comprised a total exercise time of 10.5 min, increased mean HR to levels above 80% HR_{max} for a total of 9.2 min, and above 90% HR_{max} for a total of 5.4 min (sum of time in set 1 and set 2, see Table 2). This extended time spent at a high HR may be an appropriate stimulus for improvements in aerobic fitness in very young female gymnasts. These findings are important, since circuit training using functional sport-specific exercises is routinely used in developing athletes, mainly to improve neuromuscular fitness [24]. However, exercise performed intermittently at a high intensity has been shown to improve not only strength and muscle endurance, but also to involve a significant aerobic contribution [25,26].

One interesting observation is that the time during which the heart rate was >80% HR_{max} was about 70% of the exercise plus recovery duration. Notably, these young athletes spent 9.2 min out of a total of ~16 min of exercise and recovery (2 × 5:15 min separated by 3 min or rest) with a high HR (Table 2). Furthermore, during 5.4 min of this time, gymnasts had an HR above 90% of HR_{max} (Figure 1). This time spent at a high HR is an adequate stimulus for improving VO_{2max} and aerobic fitness in general [4,27]. The fact that such a large part of this brief exercise scheme was performed with a high HR may be due to the rapid HR kinetics of children, together with their higher oxidative capacity and aerobic contribution to high-intensity exercise [8,28]. Thus, these findings provide evidence that cardiorespiratory stress is high during this type of high-intensity, sport-specific circuit training. Interestingly, a very recent study compared the acute effects of an integrative neuromuscular training program for 12 min (2 sets × 6 exercises × 30 s each with equal rest) on cardiometabolic responses of

10–11 year old children [15]. In that study, the increase of HR was lower than in the present study and HR ranged between 61% and 92% of HR_{max} [15]. Importantly, in that study, VO_2 during the 12 min exercise was increased from 28% to 64% VO_{2max}, suggesting that HR is indicative of oxygen uptake in this type of protocol. This would suggest that in the present study, more than 50% of the protocol duration was performed with high VO_2, as it is known that 90% HR_{max} corresponds to >80% of VO_{2max} [29]. Mandigout et al. [22] found that training intensities greater than 80% HR_{max} for at least 25 min per session, resulted in improved VO_{2max} in children aged 10–11 years. Thus, the present circuit exercise protocol may provide an appropriate stimulus to improve aerobic fitness only in terms of intensity (>80% HR_{max}) and not in terms of duration (i.e., a total of 9.2 min above >80% HR_{max}).

Different exercise bout configurations during intermittent functional training protocols may modify the physiological strain [2]. For example, physiological responses may vary greatly by changing the duration of work and recovery periods and this has been known for many decades [2,30]. In the present study, the work and rest durations were very brief (7 s), and as a consequence, HR and most probably VO_2, remained elevated throughout exercise, mimicking the responses during high-intensity continuous work. Longer exercise and rest durations are expected to cause a higher contribution of anaerobic glycolysis during exercises, combined with a drop in HR and VO_2 during the recovery intervals, thus reducing the cardiorespiratory strain [7,31]. Along this line, Bendiksen et al. [32] reported that mean HR and time spent in high-intensity aerobic training zones was higher in ball games (2 sets × 15 min with 3 min rest) compared to circuit resistance training (30 s work and 45 s rest for 3 min). In another study, Faigenbaum et al. [33] examined acute cardiometabolic responses, applying 10 min medicine ball (2.3 kg) interval training comprising 2 sets with 30 s work and equal rest intervals. It was found that peak HR reached 178 ± 9 bpm and that mean HR ranged from 61.1% to 81.6%. These values are lower than the values reported in the present study, probably due to the extended work and interval duration. Indeed, longer exercise durations of high-intensity intermittent exercise (15–30 s at intensities >100% VO_{2max}) are related with early exhaustion in child and adolescent athletes, and, in this case, a continuous bout of near-maximal exercise 80–90% VO_{2max} may be more effective to stimulate aerobic adaptations [21,34]. Moreover, muscle oxygenation, as measured by near-infrared resonance spectroscopy, is higher during shorter than longer duration exercise; rest intervals (24 s:36 s and 6 s:9 s, respectively) [35].

Another important finding of the present study was the rapid decrease of HR following both the functional sport-specific circuit training protocol and the shuttle run test (Figure 1, Table 2). Notably, the decrease in HR after 1 and 2 min of recovery was similar in both bouts and in the shuttle run test, suggesting that HR recovery in children is minimally affected by the characteristics of the preceding exercise bout. Previous studies have reported that post-exercise heart rate recovery is faster in children compared with adults, probably due to their lower work rate and less anaerobic metabolism contribution [36]. In a study comparing heart rate recovery between prepubertal, pubertal and adult males, after repeated high-intensity cycling sprints, it was shown that HR 1 min after exercise recovered by 50 ± 1 bpm, 37 ± 1 bpm and 39 ± 1 bpm, for the three age groups respectively, with no significant difference between adolescents and adults [37]. The data for HR recovery in children in that study [37] are similar with the findings of the present study (Table 2), demonstrating the rapid HR recovery in female gymnasts following this high-intensity circuit training routine. Possible reasons for the faster HR recovery may be a lower glycolytic energy supply coupled with a higher aerobic contribution and phosphocreatine resynthesis between bouts, as well as a greater parasympathetic reactivation [8,38,39].

In summary, this study presented novel and practically significant findings related to high-intensity sport-specific circuit training in child female gymnasts. However, there are certain limitations that should be acknowledged. Despite the fact that HR was continuously measured in the present study, VO_2 responses were not evaluated. The 20 m shuttle run test, commonly used in youth athletes, is not a sport-specific test for cardiorespiratory fitness in young gymnasts. However, there is currently no other sport-specific test to asses this fitness parameter in this population. Finally, blood lactate measurements would have been informative regarding the strain placed on anaerobic glycolysis during this high-intensity workout applied in young female gymnasts. Nevertheless, it was shown that this exercise program, that is commonly applied to enhance neuromuscular performance in young female gymnasts, is characterized by an increased heart rate, above an intensity that may induce aerobic adaptations (80% HR_{max}), albeit for a relatively short time. The time spent at high HR may be an appropriate stimulus for improvements in aerobic fitness in youth athletes. At the same time, performing different types of exercises from hanging and support on gymnastics apparatuses using body weight, may simultaneously enhance physical fitness and improve motor skills, especially in very young athletes. Further research should investigate the long-term effects of this training modality using different exercise durations on aerobic fitness, strength and power in child athletes of sports demanding high power and fast recovery abilities.

Author Contributions: Conceptualization, G.C.B. and A.S.; methodology, O.D., C.K., A.S., G.C.B.; investigation, A.S., C.K.; data analysis, A.S., G.C.B. writing—original draft preparation, A.S., O.D., C.K. writing—review and editing, A.S., O.D., C.K., G.C.B.; supervision, O.D., G.C.B. All authors have read and agreed to the published version of the manuscript.

Funding: This research received no external funding.

Acknowledgments: The authors would like to thank the young athletes and their parents.

Conflicts of Interest: The authors declare no conflict of interest.

References

1. Billat, L.V. Interval Training for Performance: A Scientific and Empirical Practice. Special Recommendations for Middle- and Long-Distance Running. Part I: Aerobic Interval Training. *Sports Med.* **2001**, *31*, 13–31. [CrossRef] [PubMed]
2. Billat, V.L. Interval Training for Performance: A Scientific and Empirical Practice: Special Recommendations for Middle- and Long-Distance Running. Part II: Anaerobic Interval Training. *Sports Med.* **2001**, 75–90. [CrossRef] [PubMed]
3. Laursen, P.B.; Jenkins, D.G. The Scientific Basis for High-Intensity Interval Training: Optimising Training Programmes and Maximising Performance in Highly Trained Endurance Athletes. *Sports Med.* **2002**, *32*, 53–73. [CrossRef] [PubMed]
4. Buchheit, M.; Laursen, P.B. High-Intensity Interval Training, Solutions to the Programming Puzzle: Part I: Cardiopulmonary Emphasis. *Sports Med.* **2013**, *43*, 313–338. [CrossRef] [PubMed]
5. Engel, F.A.; Ackermann, A.; Chtourou, H.; Sperlich, B. High-Intensity Interval Training Performed by Young Athletes: A Systematic Review and Meta-Analysis. *Front. Physiol.* **2018**, *9*. [CrossRef] [PubMed]
6. Christmass, M.A.; Dawson, B.; Goodman, C.; Arthur, P.G. Brief intense exercise followed by passive recovery modifies the pattern of fuel use in humans during subsequent sustained intermittent exercise. *Acta Physiol. Scand.* **2001**, *172*, 39–52. [CrossRef]
7. Bogdanis, G.C.; Nevill, M.E.; Boobis, L.H.; Lakomy, H.K. Contribution of Phosphocreatine and Aerobic Metabolism to Energy Supply during Repeated Sprint Exercise. *J. Appl. Physiol.* **1996**, *80*, 876–884. [CrossRef]
8. Ratel, S.; Blazevich, A.J. Are Prepubertal Children Metabolically Comparable to Well-Trained Adult Endurance Athletes? *Sports Med.* **2017**, *47*, 1477–1485. [CrossRef]
9. Eddolls, W.T.B.; McNarry, M.A.; Stratton, G.; Winn, C.O.N.; Mackintosh, K.A. High-Intensity Interval Training Interventions in Children and Adolescents: A Systematic Review. *Sports Med.* **2017**, *47*, 2363–2374. [CrossRef]
10. Baquet, G.; Gamelin, F.X.; Aucouturier, J.; Berthoin, S. Cardiorespiratory Responses to Continuous and Intermittent Exercises in Children. *Int. J. Sports Med.* **2017**, *38*, 755–762. [CrossRef]
11. Baquet, G.; Guinhouya, C.; Dupont, G.; Nourry, C.; Berthoin, S. Effects of a Short-Term Interval Training Program on Physical Fitness in Prepubertal Children. *J. Strength Cond. Res.* **2004**, 708–713. [CrossRef]
12. Greenlee, T.A.; Greene, D.R.; Ward, N.J.; Reeser, G.E.; Allen, C.M.; Baumgartner, N.W.; Cohen, N.J.; Kramer, A.F.; Barbey, A.K.; Hillman, C.H. Effectiveness of a 16-Week High-Intensity Cardio-Resistance Training (HICRT) Program in Adults. *Med. Sci. Sports Exerc.* **2016**, *48*, 860. [CrossRef]

13. Wilke, J.; Kaiser, S.; Niederer, D.; Kalo, K.; Engeroff, T.; Morath, C.; Vogt, L.; Banzer, W. Effects of High-Intensity Functional Circuit Training on Motor Function and Sport Motivation in Healthy, Inactive Adults. *Scand. J. Med. Sci. Sports* **2019**, *29*, 144–153. [CrossRef] [PubMed]
14. Moeskops, S.; Oliver, J.L.; Read, P.J.; Cronin, J.B.; Myer, G.D.; Lloyd, R.S. The Physiological Demands of Youth Artistic Gymnastics; Applications to Strength and Conditioning. *Strength Cond. J.* **2018**, *1*. [CrossRef]
15. Faigenbaum, A.D.; Kang, J.; Ratamess, N.A.; Farrell, A.C.; Belfert, M.; Duffy, S.; Jenson, C.; Bush, J. Acute Cardiometabolic Responses to Multi-Modal Integrative Neuromuscular Training in Children. *J. Funct. Morphol. Kinesiol.* **2019**, *4*, 39. [CrossRef]
16. Bogdanis, G.C.C.; Nevill, M.E.E.; Boobis, L.H.H.; Lakomy, H.K.K.A.; Nevill, A.M.M. Recovery of Power Output and Muscle Metabolites Following 30 s of Maximal Sprint Cycling in Man. *J. Physiol.* **1995**, *482*, 467–480. [CrossRef]
17. Marina, M.; Rodríguez, F.A. Physiological Demands of Young Women's Competitive Gymnastic Routines. *Biol. Sport* **2014**. [CrossRef]
18. Douda, H.T.; Toubekis, A.G.; Avloniti, A.A.; Tokmakidis, S.P. Physiological and Anthropometric Determinants of Rhythmic Gymnastics Performance. *Int. J. Sports Physiol. Perform.* **2008**, *3*, 41–54. [CrossRef]
19. Sands, W.A.; McNeal, J.R. A Minimalist Approach to Conditioning for Women's Gymnastics. In *USA Gymnastics Congress Proceedings Book*; USA Gymnastics: Indianapolis, IN, USA, 1997; pp. 78–80.
20. Léger, L.A.; Mercier, D.; Gadoury, C.; Lambert, J. The Multistage 20 Metre Shuttle Run Test for Aerobic Fitness. *J. Sports Sci.* **1988**, *6*, 93–101. [CrossRef]
21. Zafeiridis, A.; Sarivasiliou, H.; Dipla, K.; Vrabas, I.S. The Effects of Heavy Continuous versus Long and Short Intermittent Aerobic Exercise Protocols on Oxygen Consumption, Heart Rate, and Lactate Responses in Adolescents. *Eur. J. Appl. Physiol.* **2010**, *110*, 17–26. [CrossRef]
22. Mandigout, S.; Melin, A.; Lecoq, A.M.; Courteix, D.; Obert, P. Effect of Two Aerobic Training Regimens on the Cardiorespiratory Response of Prepubertal Boys and Girls. *Acta Paediatr.* **2002**, *91*, 403–408. [CrossRef] [PubMed]
23. Cohen, J. A Power Primer. *Psychol Bull.* **1992**, 155–159. [CrossRef]
24. Moeskops, S.; Read, P.; Oliver, J.; Lloyd, R. Individual Responses to an 8-Week Neuromuscular Training Intervention in Trained Pre-Pubescent Female Artistic Gymnasts. *Sports* **2018**, *6*, 128. [CrossRef] [PubMed]
25. Ratel, S.; Williams, C.A.; Oliver, J.; Armstrong, N. Effects of Age and Recovery Duration on Performance during Multiple Treadmill Sprints. *Int. J. Sports Med.* **2006**, *27*, 1–8. [CrossRef] [PubMed]
26. Bogdanis, G.C. Effects of Physical Activity and Inactivity on Muscle Fatigue. *Front. Physiol.* **2012**, *3*, 142. [CrossRef]
27. Rowland, T.W. Aerobic Response to Endurance Training in Prepubescent Children: A Critical Analysis. *Med. Sci. Sports Exerc.* **1985**, *17*, 493–497. [CrossRef]
28. Zanconato, S.; Cooper, D.M.; Armon, Y. Oxygen Cost and Oxygen Uptake Dynamics and Recovery with 1 Min of Exercise in Children and Adults. *J. Appl. Physiol.* **1991**, *71*, 993–998. [CrossRef]
29. Midgley, A.W.; Mc Naughton, L.R. Time at or near VO_2max during Continuous and Intermittent Running: A Review with Special Reference to Considerations for the Optimisation of Training Protocols to Elicit the Longest Time at or near VO_2max. *J. Sports Med. Phys. Fit.* **2006**, *46*, 1–14.
30. Åstrand, I.; Åstrand, P.-O.; Christensen, E.H.; Hedman, R. Intermittent Muscular Work. *Acta Physiol. Scand.* **1960**, *48*, 448–453. [CrossRef]
31. Bogdanis, G.C.C.; Nevill, M.E.E.; Lakomy, H.K.A.K.; Boobis, L.H.H. Power Output and Muscle Metabolism during and Following Recovery from 10 and 20 s of Maximal Sprint Exercise in Humans. *Acta Physiol. Scand.* **1998**, *163*, 261–272. [CrossRef]
32. Bendiksen, M.; Williams, C.A.; Hornstrup, T.; Clausen, H.; Kloppenborg, J.; Shumikhin, D.; Brito, J.; Horton, J.; Barene, S.; Jackman, S.R.; et al. Heart Rate Response and Fitness Effects of Various Types of Physical Education for 8- to 9-Year-Old Schoolchildren. *Eur. J. Sport Sci.* **2014**, *14*, 861–869. [CrossRef] [PubMed]
33. Faigenbaum, A.D.; Kang, J.; Ratamess, N.A.; Farrell, A.; Ellis, N.; Vought, I.; Bush, J. Acute Cardiometabolic Responses to Medicine Ball Exercise in Children. *Int. J. Exerc. Sci.* **2018**, *11*, 886–899. [CrossRef] [PubMed]
34. Berthoin, S.; Baquet, G.; Dupont, G.; Van Praagh, E. Critical Velocity during Continuous and Intermittent Exercises in Children. *Eur. J. Appl. Physiol.* **2006**, *98*, 132–138. [CrossRef] [PubMed]

35. Christmass, M.A.; Dawson, B.; Arthur, P.G. Effect of Work and Recovery Duration on Skeletal Muscle Oxygenation and Fuel Use during Sustained Intermittent Exercise. *Eur. J. Appl. Physiol. Occup. Physiol.* **1999**, *80*, 436–447. [CrossRef]
36. Birat, A.; Bourdier, P.; Piponnier, E.; Blazevich, A.J.; Maciejewski, H.; Duché, P.; Ratel, S. Metabolic and Fatigue Profiles Are Comparable Between Prepubertal Children and Well-Trained Adult Endurance Athletes. *Front. Physiol.* **2018**, *9*, 387. [CrossRef]
37. Buchheit, M.; Duché, P.; Laursen, P.B.; Ratel, S. Postexercise Heart Rate Recovery in Children: Relationship with Power Output, Blood PH, and Lactate. *Appl. Physiol. Nutr. Metab.* **2010**, *35*, 142–150. [CrossRef]
38. Tonson, A.; Ratel, S.; Le Fur, Y.; Vilmen, C.; Cozzone, P.J.; Bendahan, D. Muscle Energetics Changes throughout Maturation: A Quantitative 31P-MRS Analysis. *J. Appl. Physiol.* **2010**, *109*, 1769–1778. [CrossRef]
39. Ohuchi, H.; Suzuki, H.; Yasuda, K.; Arakaki, Y.; Echigo, S.; Kamiya, T. Heart Rate Recovery after Exercise and Cardiac Autonomic Nervous Activity in Children. *Pediatr. Res.* **2000**, *47*, 329–335. [CrossRef]

Article

Effects of Supplementary Strength–Power Training on Neuromuscular Performance in Young Female Athletes

Konstantina Karagianni, Olyvia Donti, Christos Katsikas and Gregory C. Bogdanis *

School of Physical Education and Sports Science, National and Kapodistrian University of Athens, 17237 Athens, Greece; karagiannik97@gmail.com (K.K.); odonti@phed.uoa.gr (O.D.); ckatsikas@phed.uoa.gr (C.K.)

* Correspondence: gbogdanis@phed.uoa.gr; Tel.: +30-(210)-7276115

Received: 23 June 2020; Accepted: 22 July 2020; Published: 24 July 2020

Abstract: This study examined the effects of a short-duration supplementary strength–power training program on neuromuscular performance and sport-specific skills in adolescent athletes. Twenty-three female "Gymnastics for All" athletes, aged 13 ± 2 years, were divided into a training group (TG, $n = 12$) and a control group (CG, $n = 11$). Both groups underwent a test battery before and after 10 weeks of intervention. TG completed, in addition to gymnastics training, a supplementary 7–9 min program that included two rounds of strength and power exercises for arms, torso, and legs, executed in a circuit fashion with 1 min rest between rounds, three times per week. Initially, six exercises were performed (15 s work–15 s rest), while the number of exercises was decreased to four and the duration of each exercise was increased to 30 s (30 s rest) after the fifth week. TG improved countermovement jump performance with one leg (11.5% ± 10.4%, $p = 0.002$) and two legs (8.2% ± 8.8%, $p = 0.004$), drop jump performance (14.4% ± 12.6%, $p = 0.038$), single-leg jumping agility (13.6% ± 5.2%, $p = 0.001$), and sport-specific performance (8.8% ± 7.4%, $p = 0.004$), but not 10 m sprint performance (2.4% ± 6.6%, $p = 0.709$). No change was observed in the CG ($p = 0.41$ to 0.97). The results of this study indicated that this supplementary strength–power program performed for 7–9 min improves neuromuscular and sport-specific performance after 10 weeks of training.

Keywords: female; adolescence; resistance training; plyometric training; strength training

---

1. Introduction

Long-term athlete development models provide general frameworks to prepare youth for sports and a physically active lifestyle [1]. These models aim to align practice with growth, maturation, and early sport specialization and to consider factors such as injury risk [2,3] and the limitations of the existing training practice schedules [3,4]. Muscular strength and power, speed, and agility are central fitness components in all long-term athlete development models, which propose participation in high-intensity physical activity and muscle strengthening exercises at least 3 times a week [3,5]. Muscular strength and power increase with age in boys and girls until the onset of puberty [6,7], while after this age, a plateau in muscular strength is typically observed in girls [8]. Sprint speed increases in early childhood (5–9 years) and shows accelerated improvement in boys ~12 to 14 years; however, improvements occur earlier in girls, and sprint performance reaches a plateau 2 to 3 years earlier than boys [9,10].

Resistance and plyometric training are efficient methods to improve strength, power, speed, and general athleticism in youth athletes [11,12]. Importantly, enhanced physical fitness is a prerequisite for motor competence and technical skill acquisition, in youth athletes [1]. Recently, Lesinski et al. [13] demonstrated the positive effects of resistance training (including weight-bearing exercises) on muscular strength and jumping performance in youth athletes, and especially in adolescents, with boys improving more than girls [13,14]. Plyometric training is also an effective method to increase lower-limb strength and power, sprinting, and change of direction abilities in young male athletes [15,16]. However, research is lacking in female athletes and in individual sports [17]. In a recent meta-analysis examining the effect of plyometric training in female athletes (8–18 years), Moran et al. [18] reported small-to-moderate effects on jumping performance, with larger effect sizes observed in younger (<15 years; ES = 0.78) compared to older athletes (>15 years; ES = 0.31). Nevertheless, limited research examined the extent to which young female athletes increase strength and power following different modes of training. This would be especially useful in sports such as gymnastics, where girls train and compete from a very early age [19], and increased physical fitness is associated not only with sports performance, but also with reduced injury occurrence.

"Gymnastics for All" is an early specialization sport, which incorporates various elements (artistic, rhythmic, acrobatic, and aerobic) executed by a group of athletes on the gymnastics floor. Performance of these complex skills requires high levels of relative strength, power, and flexibility [19–22], while aerobic capacity has also been found important for both competition scores and recovery [23–25] throughout training and competition. Due to the less competitive character of "Gymnastics for All" (group contest or festival), athletes typically train 3–4 times per week for 1.5–2 h per session. Gymnastics coaches often use combinations of skills repetitions in a circuit type of training, to improve athletes' neuromuscular fitness [26,27]. However, while this training is commonly used in gymnastics to develop sport-specific fitness, the addition of an age appropriate, supplementary, strength and power training program may offer positive outcomes that surpass the benefits obtained by skills training alone [26,28]. Previous studies suggested that supplementary strength and power training using basic movement skills, may enhance technical competency, correct movement patterns, and reduce injury risk [26,29] especially in adolescent female athletes, who show decreased strength and increased injury risk compared to males [30,31]. An important limitation of the existing training practice schedules in most youth sports is that training time per session (typically 1.5 h) may not be sufficient for learning new skills and improving muscular fitness. However, there is limited information on possible modifications needed in such training schedules in order to conform to the guidelines of current long-term athlete development models [1]. For example, Moeskops et al. [26] used two weekly neuromuscular training sessions that lasted 35 min each, during an 8 week period and found increased leg stiffness and muscular endurance in 8–9-year-old gymnasts. However, such durations of fitness training may not be feasible in most youth sport training schedules, due to time restrictions. Thus, it may be useful to examine shorter-duration fitness interventions.

Recently, it was suggested that a combination of two or more types of training may be more efficient to improve fitness and sport performance of youth athletes [32–34]. For example, a combined resistance and plyometric training program, performed for 6 weeks enhanced maximal strength, countermovement jump height, and sprint speed in three groups of youth basketball players (13–15, 15–17, and >17 years), although the training program used in that study was less effective as the age of the basketball players increased [32]. However, the effects of such combined strength and power programs in younger athletes are still unclear. Thus, the aim of this study was to examine the effects of a 10 week, short-duration, supplementary strength–power training program on neuromuscular performance and sport-specific skills in female, adolescent "Gymnastics for All" athletes. It was hypothesized that this supplementary strength–power training would improve physical fitness and in turn performance in sport-specific skills, more than gymnastics training alone.

2. Materials and Methods

2.1. Participants

Power analysis indicated that a minimum of 6 gymnasts should be included in the study in order to detect an effect size (ES) of 0.78, obtained from the meta-analysis of Moran et al. [18] for girls younger than 15 years of age (within–between analysis of variance power = 0.80, alpha = 0.05, correlation between repeated measures $r = 0.5$; G-Power 3.1.9.2).

Twenty-six female "Gymnastics for All" gymnasts, aged 13 ± 2 years were recruited from one gymnastics club. Criteria for inclusion were: regular training (i.e., three times per week for 90 min per session) for at least two years under the same coach; no involvement in any systematic strength and power training; competitive experience of at least one year. Participants were excluded if they had any musculoskeletal injury in the last six months or if they missed >10% of the training sessions. Gymnasts were randomly allocated (allocation ratio 1:1) to a training group (TG) and a control group (CG). Three athletes (two from the training group and one from the control group) were excluded from the study because they did not complete all tests. The final number of participants along with their anthropometric and maturity characteristics are presented in Table 1. The maturity offset was estimated according to the prediction equation of Mirwald et al. [35]. Before the start of the study, researchers informed coaches, athletes, and their parents about the purpose and risks of the study, and an informed consent was signed by the athletes and the participants' parents. All procedures were approved by the local university ethics committee (reference number: 115/10-04-2019) in compliance with the Code of Ethics of the World Medical Association (Helsinki declaration of 1964, as revised in 2013).

Table 1. Characteristics of the participants in the training group (TG) and the control group (CG) (mean ± SD).

Characteristics	TG (n = 12)	CG (n = 11)	p
Age (year)	13.2 ± 1.3	12.3 ± 1.3	0.106
Training experience (year)	4.4 ± 2.7	4.3 ± 2.1	0.883
Height (cm)	157.3 ± 6.1	156.0 ± 7.3	0.638
Body mass (kg)	52.4 ± 6.6	51.6 ± 8.5	0.792
BMI (kg/m^2)	21.1 ± 1.6	21.2 ± 3.0	0.956
Maturity offset	1.1 ± 0.9	0.6 ± 0.9	0.159

2.2. Study Design

A repeated-measures parallel group design was used in the present study. The same battery of tests was evaluated in both groups at the beginning and at the end of the intervention (10 weeks). Athletes in the TG performed a circuit-type strength and power training program (duration: 7–9 min) for 10 weeks in addition to their regular gymnastics training. The study took place in the pre-season period (from October to December). The following tests were repeated at baseline and after 10 weeks of training: 10 m linear sprint speed, one-leg (one-leg CMJ) and two-legs countermovement jump (CMJ), drop jump (DJ), single-leg jumping agility test (JA), and 10 consecutive repetitions of a sport-specific skill (round-off). All measurements were performed in the same testing session, 48 h after the last training. A standardized warm-up preceded testing that included 6 min of light jogging, dynamic stretching for the major muscle groups, and 2 short accelerations.

Twenty-six sessions of this circuit strength and power training program were performed on non-consecutive days (Monday, Wednesday, and Friday), at the end of regular gymnastics training. Four familiarization sessions were performed over a two-week period, before baseline testing, to get participants habituated to the circuit strength-power training and the testing procedures. Both training and control groups participated in all familiarization sessions. During that period, data were collected to calculate intraclass correlation coefficients (ICCs) for each test, using a two-way mixed-model analysis of variance (ANOVA).

2.3. Methodology

2.3.1. Anthropometry

Body mass (to within 0.1 kg, Seca 700, Seca Ltd., Birmingham, UK), standing height, and sitting height (to within 0.1 cm, Charder HM-200P, Charder Electronic Co., Ltd., Taichung City 412, Taiwan) were measured for both groups.

2.3.2. Vertical Jump Height

Vertical jump height was determined from flight time using an Optojump system (Microgate, SRL, Bolzano, Italy) [36]. Participants were instructed to jump at maximal effort with their hands akimbo, take off with the ankles and knees fully extended and land balanced on the same spot. During all jumping tests participants wore gymnastics shoes. For the one- and the two-leg CMJ, gymnasts were instructed to perform a countermovement (knee angle approx. 90°) and then immediately jump up. Trials were separated by 30 s and there was a 2 min rest between the different jump tests. For the single-leg CMJ test, the sum of the right- and left-leg jumps was calculated and used for further analysis. ICC for the two-leg CMJ was 0.93 (95% confidence interval (CI): 0.86–0.97) (standard error of measurement (SEM) = 3.3%, meaningful detectable change at 90% confidence interval (MDC_{90}) = 2.17 cm) and for the sum of the right- and left-leg CMJ it was 0.94 (95% CI: 0.86–0.97) (SEM = 3.6%, MDC_{90} = 2.49 cm).

For the DJ, gymnasts stepped horizontally off a 20 cm box on the gymnastics carpet and then immediately performed a maximal rebound vertical jump with minimal ground contact time. The ICC for the DJ height was 0.98 (95% CI: 0.96–0.99) (SEM = 1.6%, MDC_{90} = 1.00 cm).

2.3.3. Single-Leg Jumping Agility Test

As for the single-leg jumping agility, a cross hop was used. This test examines jumping agility, as it requires the participants to hop as fast as possible and to move in multiple directions [37]. Subjects had to perform five consecutive rounds of hopping for each leg as fast as they could. For this test, a cross consisting of five equal-sized squares (30 cm side) was marked on the gymnastics floor with tape. The distance of the front, back, right, and left square from the middle (central) square, was set at 20% of the participant's body height, measured from the center of the middle square, to the center of all the other squares. The starting point of the test was the central square, where gymnast stood on one leg, and from there they had to jump forward, backward, to the right. and to the left, into the respective squares, always returning to the central square after each jump (one round). The total time to complete the 5 rounds was recorded electronically. When a gymnast touched the contralateral foot on the ground, or hopped in the wrong direction, they had to repeat the trial. Two trials interspersed by 5 min of rest were performed, and the best trial was used for further analysis. The average value of the right- and left-leg performance was calculated for further analysis. The ICC for single-leg jumping agility was 0.97 (95% CI: 0.93–0.98) (SEM = 1%, MDC_{90} = 0.62 s).

2.3.4. Sprint Test

Ten meter linear sprint performance was assessed electronically (Microgate, SARL, Bolzano, Italy) [36]. Participants were asked to stand in an upright stride stance with the preferred leg forward, 0.3 m before the first infrared photoelectric gate, which was placed 0.75 m above the ground to ensure it captured trunk movement and avoided false limb motion signals. Two trials interspersed by 5 min of rest were performed, and the best trial was used for further analysis. The intraclass correlation coefficient for the 10 m sprint was 0.99 (95% CI: 0.98–0.99) (SEM = 0.3%, MDC_{90} = 0.015 s).

2.3.5. Sport-Specific Skill

Round-off is a basic movement in gymnastics used to gain speed before performing a series of flic-flacs and saltos [38]. The round-off includes the following phases: (a) a run-up phase, which ends by placing the hands on the floor in a T-shape, while inverting the body; (b) the main phase of support, followed by a rapid push-off from the hands and snap down; and (c) the last phase in which the feet land together on the floor while the body is inverted [38]. Participants performed 10 consecutive repetitions of round-off as fast as possible, each one starting from two steps. The total time to complete the 10 repetitions was measured electronically. Two trials interspersed by 5 min of rest were performed, and the best trial was used for further analysis. The intraclass correlation coefficient for the 10 round-offs was 0.99 (95% CI: 0.98–0.99) (SEM = 0.7%, MDC_{90} = 0.42 s).

2.3.6. Strength and Power Training

At the end of the training session (Monday, Wednesday, and Friday), gymnasts of the TG performed a circuit strength and power training program while at the same time the athletes of the CG performed body posture movements. This circuit strength and power program included two 5 week training blocks. Six strength and power exercises of progressive difficulty for arms, torso, and legs were performed in the first training block (15 s work–15 s rest), while the number of exercises was decreased to four and the duration of each exercise was increased to 30 s (30 s rest) after the fifth week (Table 2). Each training included a combination of strength and power exercises targeting major muscle groups (Table 2). The exercises included are presented in Table S1. Athletes were instructed to perform as many repetitions as possible during the time available for each exercise. Strength and power exercises were performed on the surface of a gymnastics carpet with the gymnasts wearing gymnastics shoes. Training was supervised by an experienced coach, and proper technique of movement was emphasized at every training and testing session.

2.4. Statistical Analyses

Descriptive statistics were calculated for all performance and anthropometric tests. The normality of data distribution and homogeneity of variance were checked using Shapiro–Wilk test and Levene's test, respectively. Unpaired t-tests were applied to determine significant differences in baseline values between groups. A two-way analysis of variance (ANOVA) [group (training/control) × time (pre/post-training)], with repeated measures on time, was conducted to examine the effect of strength and power training on all the examined parameters. When a significant main effect or interaction was observed ($p < 0.05$), a Tukey's post hoc test was performed. Effect sizes (ES) for the ANOVA were determined by partial eta squared (η^2). Partial eta squared (η^2) values were classified as small (0.01 to 0.059), moderate (0.06 to 0.137), and large (>0.137). For pairwise comparisons, ES was determined by Cohen's *d* [39] (small: >0.2, medium: >0.5, and large: >0.80). The intraclass correlation coefficient (ICC) was calculated using a two-way mixed model, to measure reliability for all measures. Additionally, the standard error of measurement (SEM) and the meaningful detectable change at 90% confidence interval (MDC_{90}) were calculated. Statistical significance was set at $p < 0.05$. All statistical analyses were conducted using SPSS (IBM SPSS Statistics, Version 22.0, IBM Corporation, Armonk, New York, USA).

Table 2. Strength and power training program executed three times per week for 10 weeks by the athletes of the training group (TG). Athletes performed two rounds of six exercises in a circuit form, for the first 6 weeks and two rounds of four exercises for the last 4 weeks.

	MONDAY	WEDNESDAY	FRIDAY
WEEK 1	2 × 6 exercises (5 S + 1 P) W:R = 15:15 total duration: 7 min	2 × 6 exercises (5 S + 1 P) W:R = 15:15 total duration: 7 min	2 × 6 exercises (5 S + 1 P) W:R = 15:15 total duration: 7 min
WEEK 2	2 × 6 exercises (2 S + 4 P) W:R = 15:15 total duration: 7 min		2 × 6 exercises (4 S + 2 P) W:R = 15:15 total duration: 7 min
WEEK 3	2 × 6 exercises (1 S + 5 P) W:R = 15:15 total duration: 7 min	2 × 6 exercises (3 S + 3 P) W:R = 15:15 total duration: 7 min	2 × 6 exercises (1 S + 5 P) W:R = 15:15 total duration: 7 min
WEEK 4	2 × 6 exercises (1 S + 5 P) W:R = 15:15 total duration: 7 min	2 × 6 exercises (4 S + 2 P) W:R = 15:15 total duration: 7 min	2 × 6 exercises (4 S + 2 P) W:R = 15:15 total duration: 7 min
WEEK 5	2 × 6 exercises (6 P) W:R = 15:15 total duration: 7 min		2 × 6 exercises (1 S + 5 P) W:R = 15:15 total duration: 7 min
WEEK 6	2 × 6 exercises (1 S + 5 P) W:R = 20:20 total duration: 9 min	2 × 6 exercises (2 S + 4 P) W:R = 20:20 total duration: 9 min	2 × 6 exercises (1 S + 5 P) W:R = 20:20 total duration: 9 min
WEEK 7	2 × 4 exercises (2 S + 2 P) W:R = 30:30 total duration: 9 min	2 × 4 exercises (1 S + 3 P) W:R = 30:30 total duration: 9 min	2 × 4 exercises (2 S + 2 P) W:R = 30:30 total duration: 9 min
WEEK 8	2 × 4 exercises (2 S + 2 P) W:R = 30:30 total duration: 9 min		2 × 4 exercises (4 S) W:R = 30:30 total duration: 9 min
WEEK 9	2 × 4 exercises (4 S) W:R = 30:30 total duration: 9 min	2 × 4 exercises (1 S + 3 P) W:R = 30:30 total duration: 9 min	2 × 4 exercises (4 S) W:R = 30:30 total duration: 9 min
WEEK 10	2 × 4 exercises (4 S) W:R = 30:30 total duration: 9 min		2 × 4 exercises (4 P) W:R = 30:30 total duration: 9 min

S: strength exercises, P: plyometric exercises, W:R: work-to-rest ratio.

3. Results

Performance Parameters

No statistical differences were observed between groups in baseline values (Table 1). Group × time interactions were found for all the examined parameters ($p < 0.038$) except for 10 m sprint speed ($p = 0.709$) (Table 3). Comparison of changes in performance values between the TG and CG showed large effect sizes, indicating improvement in one- and two-legs countermovement jumps, drop jump, single-leg jumping agility, and round-off performance time in the TG (Table 3).

In particular, significant group × time interactions were observed for two- and one-leg countermovement jumps height ($p = 0.004$, $\eta^2 = 0.329$ and $p = 0.002$, $\eta^2 = 0.351$, respectively) with the post hoc showing an improvement only for the TG ($p = 0.032$, $d = 1.20$ and $p = 0.003$, $d = 0.73$, respectively) (Table 3). Significant group × time interactions were also observed in round-off performance time and single-leg jumping agility ($p = 0.004$, $\eta^2 = 0.328$ and $p = 0.001$, $\eta^2 = 0.398$, respectively) with the post hoc test showing improvement only for the TG ($p = 0.0109$, $d = 1.20$ and $p = 0.001$, $d = 2.15$). A significant group × time interaction was also observed for drop-jump ($p = 0.038$, $\eta^2 = 0.189$) with the post hoc test showing improvement only or the TG ($p = 0.008$, $d = 0.98$).

Table 3. Changes in the examined parameters following 10 weeks of intervention in the training group (TG, n = 12) and the control group (CG, n = 11).

Measured Parameter	Group	Pre-Training	Post-Training	p (Interaction)	Cohen's d (Pre vs. Post)	Δ Values (Pre vs. Post)	Cohens' d of Δ Values Between Groups
10 m Sprint (s)	TG	2.05 ± 0.10	2.09 ± 0.11	0.709	0.44	0.04 ± 0.13	0.17
	CG	2.14 ± 0.07	2.17 ± 0.09		0.35	0.03 ± 0.08	
Round-Off (s)	TG	23.17 ± 2.56	20.97 ± 0.91	0.004	1.20	2.21 ± 2.00	1.47
	CG	23.09 ± 1.97	23.81 ± 1.86		0.40	0.72 ± 2.18	
CMJ (cm)	TG	24.00 ± 3.34	26.03 ± 4.68		0.52	2.0 ± 2.27	1.47
	CG	22.95 ± 3.66	21.82 ± 3.66	0.004	0.32	1.12 ± 2.22	
R + L CMJ (cm)	TG	24.99 ± 3.64	27.81 ± 4.40		0.73	2.82 ± 2.49	1.54
	CG	22.79 ± 3.31	22.20 ± 3.11	0.002	0.20	0.6 ± 2.10	
DJ (cm)	TG	22.12 ± 2.55	25.33 ± 4.15		0.98	3.21 ± 2.77	1.01
	CG	19.93 ± 3.98	20.30 ± 2.00	0.038	0.12	0.37 ± 3.10	
Single-leg Jumping Agility (s)	TG	17.91 ± 1.48	15.43 ± 0.86		2.15	2.48 ± 1.03	1.70
	CG	18.62 ± 1.30	18.09 ± 1.38	0.001	0.42	0.53 ± 1.36	

Note: CMJ: counter movement jump; R + L CMJ: sum of the right- and the left-leg counter movement jumps; DJ: drop jump.

The percent changes in performance for the TG and the CG are presented in Figure 1. TG improved CMJ performance with one-leg (11.5% ± 10.4%, p = 0.002) and two-legs (8.2% ± 8.8%, p = 0.004), DJ performance (14.4% ± 12.6%, p = 0.038), single-leg jumping agility (13.6% ± 5.2%, p = 0.001), and sport-specific performance (8.8% ± 7.4%, p = 0.004), but not 10 m sprint performance (2.4% ± 6.6%, p = 0.709). All performance changes for the CG were not significant (p = 0.41 to 0.98).

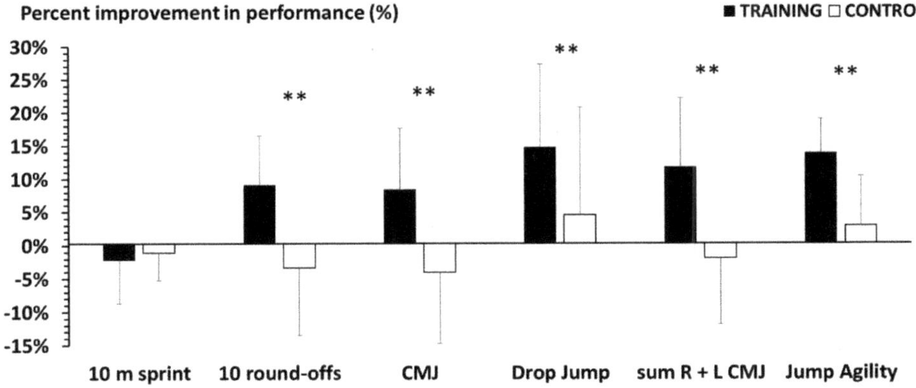

Figure 1. Percentage of change pre–post intervention in the examined parameters in the training and the control groups; CMJ: counter movement jump; R + L CMJ: sum of right– and the left–leg counter movement jumps; DJ: drop jump; ** $p < 0.01$ between training and control groups.

4. Discussion

The main finding of this study was that a 10 week circuit-type strength and power training program of short duration (7–9 min) was effective in increasing jumping height, single-leg jumping agility, and sport-specific skill performance time in adolescent female gymnasts. Ten meter sprint speed remained unchanged in both groups, suggesting that this training program, as well as gymnastics training alone, did not improve speed abilities in adolescent female gymnasts. To the authors' knowledge, this is the first study to examine a short-duration combined strength and power training program on jumping abilities, sprinting, and sport-specific parameters in young adolescent gymnasts. Importantly, the participants of this study were younger than 15 years, which is the age at which full adult height is typically achieved in females, thus representing a population undergoing maturational change [40].

The fact that a very short-duration (7–9 min) circuit-type program aiming to enhance whole-body strength and power training was so effective in improving jumping abilities, single-leg jumping agility, and sport-specific skill performance time in adolescent gymnasts, is of high practical value. This is because in most youth sports, athletes train 2–4 times per week for a limited time on each session (usually 1.5 h). During this restricted training time, coaches should allocate the necessary time to improve technical skills, team tactics, and physical fitness. Although previous research has shown that plyometric or strength and power training programs of longer duration (25 to 45 min) enhanced lower-limb strength and power in competitive gymnasts [26,41,42], such fitness training duration may not be feasible in most popular child sports, where technical and tactical training is prioritized. Taking this into account, a short-duration, whole-body strength and power program, such as the one used in the present study, would offer an efficient way to improve strength, power, and sport-specific skills, with minimal time investment.

The magnitude of improvements in single- and double-leg CMJ and in the DJ (Table 3, Figure 1), with effect sizes between 0.52 and 0.98, are in line with the findings of Moran et al. [18], who reported that vertical jump ability is developed to a greater degree in younger (<15 years) than in older (>15 years) female athletes ($d = 0.78$ and $d = 0.31$, respectively). When comparing the improvement of vertical jump between the TG and the CG in the present study, the effect sizes were large (1.01 to 1.47), indicating the effectiveness of this short-duration training program. Notably, this program improved not only general but also sport-specific fitness. In contrast, typical gymnastics training of the same weekly frequency and duration (3 times per week for 1.5 h per session—control group) does not seem to be an adequate stimulus for lower-limb power and sport-specific agility in adolescent athletes. This finding of the present study is important because it demonstrates the beneficial effects of a short-duration, supplementary, strength and power exercise program in these young athletes.

Single-leg jumping agility in this study was measured with a test that requires fast hopping forward, backward, and in the sagittal plane [37]. Single-limb hopping tests are typically testing functional ankle instability and are also classified as agility maneuvers due to the sudden direction changes that are required [37]. The ankle joint is often injured in gymnastics due to rapid take-offs and landings from different heights [43]. Several previous studies examined the role of strength and power training as a means of injury prevention in female athletes [44–46], but only a limited number of studies examined this relationship in female youth. Moran et al. [18] argued that a higher level of physical fitness not only increases performance but may also offset injury risk in child athletes. The strength and power program implemented in this study included exercises for arms, legs, and torso strength and power, thus, enhancing balance, coordination, and speed which would, in turn, improve single-leg jumping agility [34]. The pattern of large between-group differences was also evident in this test (between groups, $d = 1.70$) showing that sport-specific-skills repetition alone is not an adequate stimulus to improve a functional test associated with injury variables.

A running speed of at least 6.20 $m{\cdot}s^{-1}$, attained during 20 m run-up, is required for effective execution of a stretched salto [26,47]. Thus, the ability to accelerate, as well as the ability to execute fast technical "transition skills", such as the round-off, is decisive for successful and safe performance of acrobatic elements in gymnastics. In the present study, athletes in the TG improved the time of execution of the round-off, possibly due to improved strength and power of the upper and lower limbs, as well as of the torso [42]. This would suggest that this brief strength and power training induces adaptations that are transferred to sport-specific skills, and thus it may be recommended for young female athletes of this sport. However, the ability to accelerate, as reflected by the 10 m sprint performance, remained unchanged in both TG and CG (Table 3). This finding is in contrast with previous research in younger female gymnasts (8–9 years old), which showed a significant but moderate improvement in 10 m performance (d = 0.40) after 8 weeks of plyometric training [41]. One possible explanation for this discrepancy may be a difference in trainability between these two age groups of athletes. In the present study, female athletes' age was between 11 and 15 years, while in that previous study [41] female gymnasts were 8–10 years old. In child female athletes, the age before growth spurt (i.e., before the age of 11 years) is known as a "window of opportunity" for speed training [4,9], while a plateau in the trainability of sprint speed is observed in females at the ages of 11–14 [9,10], along with a concomitant increase in height and body mass. Thus, it is possible that an improvement in sprint performance may be more difficult to attain in adolescents between 11 and 15 years of age. Alternatively, the lack of improvement of 10 m sprint time may suggest that this low training volume is not enough to improve sprint performance over short distances, which require repetitive powerful muscle actions [48]. One limitation of this study is that performance in longer sprint distances was not examined. Therefore, it cannot be excluded that longer sprint performance (e.g., 20 m) would have been improved, as previous studies have found a twofold greater increase in 20 m compared to 10 m sprint performance following plyometric training (d = 0.40 vs. 0.81, respectively) [41]. Another limitation of this study was that in the sport-specific skill, only the time of execution was analysed. Further research should also analyse technical parameters of sport-specific skills to examine the transfer of improved strength and power on technique. In addition, future research should consider longer training interventions and in different periods over the year to examine strength and power adaptations in adolescent athletes, including males who may have different responses compared with female athletes at this age.

5. Conclusions

In conclusion, a brief-duration, supplementary, circuit-type strength and power training program was effective in increasing single- and double-leg vertical jump, jumping agility, and sport-specific skill performance in adolescent female gymnasts. Improvement of performance by devoting only 7–9 min per session may be an attractive model of supplementary training in female adolescent athletes, which may also reduce injury risk. The fact that gymnastics skills training alone was not adequate to improve important fitness and sport-specific parameters, may indicate that a supplementary strength and power training program is necessary for an effective and safe athlete development in this age group. Practitioners should consider incorporating a short-duration, supplementary strength–power program in "Gymnastics for All" training, as it was shown that substantial gains in neuromuscular performance can be obtained with minimum time investment.

Supplementary Materials: The following are available online at http://www.mdpi.com/2075-4663/8/8/104/s1, Table S1: Plyometric and strength exercises used by the athletes of the training group.

Author Contributions: Conceptualization, G.C.B. and K.K.; methodology, G.C.B., O.D., C.K., and K.K.; investigation, K.K.; data analysis, G.C.B. and K.K.; writing—original draft preparation, O.D. and K.K.; writing—review and editing, G.C.B., O.D., C.K., and K.K.; supervision, G.C.B. and O.D. All authors have read and agreed to the published version of the manuscript.

Funding: This research received no external funding.

Acknowledgments: The authors would like to thank the young athletes and their parents.

Conflicts of Interest: The authors declare no conflict of interest.

References

1. Pichardo, A.W.; Oliver, J.L.; Harrison, C.B.; Maulder, P.S.; Lloyd, R.S. Integrating models of long-term athletic development to maximize the physical development of youth. *Int. J. Sports Sci. Coach.* **2018**, *13*, 1189–1199. [CrossRef]

2. Smith, J.J.; Eather, N.; Morgan, P.J.; Plotnikoff, R.C.; Faigenbaum, A.D.; Lubans, D.R. The health benefits of muscular fitness for children and adolescents: A systematic review and meta-analysis. *Sports Med.* **2014**, *9*, 1209–1223. [CrossRef] [PubMed]
3. Lloyd, R.S.; Oliver, J.L.; Faigenbaum, A.D.; Howard, R.; De Ste Croix, M.B.A.; Williams, C.A.; Best, T.M.; Alvar, B.A.; Micheli, L.J.; Thomas, D.P.; et al. Long-term athletic development, Part 2: Barriers to success and potential solutions. *J. Strength Cond. Res.* **2015**, *29*, 1451–1464. [CrossRef] [PubMed]
4. Lloyd, R.S.; Oliver, J.L.; Faigenbaum, A.D.; Howard, R.; De Ste Croix, M.B.A.; Williams, C.A.; Best, T.M.; Alvar, B.A.; Micheli, L.J.; Thomas, D.P.; et al. Long-term athletic development, Part 1: A pathway for all youth. *J. Strength Cond. Res.* **2015**, *29*, 1439–1450. [CrossRef] [PubMed]
5. World Health Organization (WHO). *Global Recommendations on Physical Activity for Health*; World Health Organization: Geneva, Switzerland, 2010; pp. 15–33.
6. Beunen, G.; Thomis, M. Muscular strength development in children and adolescents. *Pediatr. Exerc. Sci.* **2000**, *12*, 174–197. [CrossRef]
7. Faigenbaum, A.D.; Myer, G.D. Resistance training among young athletes: Safety, efficacy and injury prevention effects. *Br. J. Sports Med.* **2010**, *44*, 56–63. [CrossRef]
8. Malina, R.M.; Bouchard, C.; Bar-Or, O. *Growth, Maturation, and Physical Activity*, 2nd ed.; Human Kinetics: Champaign, IL, USA, 2004.
9. Papaiakovou, G.; Giannakos, A.; Michailidis, C.; Patikas, D.; Bassa, E.; Kalopisis, V.; Anthrakidis, N.; Kotzamanidis, C. The effect of chronological age and gender on the development of sprint performance during childhood and puberty. *J. Strength Cond. Res.* **2009**, *23*, 2568–2573. [CrossRef]
10. Meyers, R.W.; Oliver, J.L.; Hughes, M.G.; Lloyd, R.S.; Cronin, J.B. New insights into the development of maximal sprint speed in male youth. *Strength Cond. J.* **2017**, *39*, 2–10. [CrossRef]
11. Granacher, U.; Goesele, A.; Roggo, K.; Wischer, T.; Fischer, S.; Zuerny, C.; Gollhofer, A.; Kriemler, S. Effects and mechanisms of strength training in children. *Int. J. Sports Med.* **2011**, *32*, 357–364. [CrossRef]
12. Moran, J.J.; Sandercock, G.R.H.; Ramírez-Campillo, R.; Meylan, C.M.P.; Collison, J.A.; Parry, D.A. Age-related variation in male youth athletes' countermovement jump after plyometric training: A meta-analysis of controlled trials. *J. Strength Cond. Res.* **2017**, *31*, 552–565. [CrossRef]
13. Lesinski, M.; Prieske, O.; Granacher, U. Effects and dose-response relationships of resistance training on physical performance in youth athletes: A systematic review and meta-analysis. *Br. J. Sports Med.* **2016**, *50*, 781–795. [CrossRef]
14. Granacher, U.; Lesinski, M.; Büsch, D.; Muehlbauer, T.; Prieske, O.; Puta, C.; Gollhofer, A.; Behm, D.G. Effects of resistance training in youth athletes on muscular fitness and athletic performance: A conceptual model for long-term athlete development. *Front. Physiol.* **2016**, *7*, 164. [CrossRef] [PubMed]
15. Ramirez-Campillo, R.; Moran, J.; Chaabene, H.; Granacher, U.; Behm, D.G.; García-Hermoso, A.; Izquierdo, M. Methodological characteristics and future directions for plyometric jump training research: A scoping review update. *Scand. J. Med. Sci. Sports* **2020**, *30*, 983–997. [CrossRef] [PubMed]
16. Asadi, A.; Arazi, H.; Ramirez-Campillo, R.; Moran, J.; Izquierdo, M. Influence of Maturation Stage on Agility Performance Gains after Plyometric Training: A Systematic Review and Meta-analysis. *J. Strength Cond. Res.* **2017**, *31*, 2609–2617. [CrossRef] [PubMed]
17. Ramirez-Campillo, R.; Álvarez, C.; García-Hermoso, A.; Ramírez-Vélez, R.; Gentil, P.; Asadi, A.; Chaabene, H.; Moran, J.; Meylan, C.; García-de-Alcaraz, A.; et al. Methodological Characteristics and Future Directions for Plyometric Jump Training Research: A Scoping Review. *Sports Med.* **2018**, *48*, 1059–1081. [CrossRef]
18. Moran, J.; Clark, C.C.T.; Ramirez-Campillo, R.; Davies, M.J.; Drury, B. A Meta-Analysis of Plyometric Training in Female Youth. *J. Strength Cond. Res.* **2019**, *33*, 1996–2008. [CrossRef]
19. Sands, W.A.; McNeal, J.R.A. Minimalist Approach to Conditioning for Women's Gymnastics. In *USA Gymnastics Congress Proceedings Book*; USA Gymnastics: Indianapolis, IN, USA, 1997; pp. 78–80.
20. Donti, O.; Panidis, I.; Terzis, G.; Bogdanis, G.C. Gastrocnemius Medialis Architectural Properties at Rest and During Stretching in Female Athletes with Different Flexibility Training Background. *Sports* **2019**, *7*, 39. [CrossRef]
21. Panidi, I.; Bogdanis, G.C.; Gaspari, V.; Spiliopoulou, P.; Donti, A.; Terzis, G.; Donti, O. Gastrocnemius Medialis Architectural Properties in Flexibility Trained and Not Trained Child Female Athletes: A Pilot Study. *Sports* **2020**, *8*, 29. [CrossRef]

22. Donti, O.; Donti, A.; Theodorakou, K. A review on the changes of the evaluation system affecting artistic gymnasts' basic preparation: The aspect of choreography preparation. *Sci. Gymnast. J.* **2014**, *2*, 63–73.
23. Douda, H.T.; Toubekis, A.G.; Avloniti, A.A.; Tokmakidis, S.P. Physiological and anthropometric determinants of rhythmic gymnastics performance. *Int. J. Sports Physiol. Perform.* **2008**, *3*, 41–54. [CrossRef]
24. Bogdanis, G.C.; Nevill, M.E.; Boobis, L.H.; Lakomy, H.K.; Nevill, A.M. Recovery of power output and muscle metabolites following 30 s of maximal sprint cycling in man. *J. Physiol.* **1995**, *482*, 467–480. [CrossRef] [PubMed]
25. Bogdanis, G.C.; Nevill, M.E.; Boobis, L.H.; Lakomy, H.K.; Nevill, A.M. Contribution of phosphocreatine and aerobic metabolism to energy supply during repeated sprint exercise. *J. Appl. Physiol.* **1996**, *80*, 876–884. [CrossRef] [PubMed]
26. Moeskops, S.; Read, P.; Oliver, J.; Lloyd, R. Individual Responses to an 8-Week Neuromuscular Training Intervention in Trained Pre-Pubescent Female Artistic Gymnasts. *Sports* **2018**, *6*, 128. [CrossRef] [PubMed]
27. Salagas, A.; Donti, O.; Katsikas, C.; Bogdanis, G.C. Heart Rate Responses during Sport-Specific High-Intensity Circuit Exercise in Child Female Gymnasts. *Sports.* **2020**, *8*, 68. [CrossRef]
28. Marina, M.; Jemni, M. Plyometric training performance in elite oriented prepubertal female gymnasts. *J. Strength Cond. Res.* **2014**, *28*, 1015–1025. [CrossRef]
29. Moeskops, S.; Oliver, J.L.; Read, P.J.; Cronin, J.B.; Myer, G.D.; Lloyd, R.S. The Physiological Demands of Youth Artistic Gymnastics. *Strength Cond. J.* **2019**, *41*, 1–13. [CrossRef]
30. Hägglund, M.; Atroshi, I.; Wagner, P.; Waldén, M. Superior compliance with a neuromuscular training programme is associated with fewer ACL injuries and fewer acute knee injuries in female adolescent football players: Secondary analysis of an RCT. *Br. J. Sports Med.* **2013**, *47*, 974–979. [CrossRef]
31. Zwolski, C.; Quatman-Yates, C.; Paterno, M.V. Resistance Training in Youth: Laying the Foundation for Injury Prevention and Physical Literacy. *Sports Health* **2017**, *9*, 436–443. [CrossRef]
32. Yáñez-García, J.M.; Rodríguez-Rosell, D.; Mora-Custodio, R.; González-Badillo, J.J. Changes in Muscle Strength, Jump, and Sprint Performance in Young Elite Basketball Players: The Impact of Combined High-Speed Resistance Training and Plyometrics. *J. Strength Cond. Res.* **2019**. Available online: https://pubmed.ncbi.nlm.nih.gov/31895288/ (accessed on 27 December 2019). [CrossRef]
33. Peitz, M.; Behringer, M.; Granacher, U. A systematic review on the effects of resistance and plyometric training on physical fitness in youth- What do comparative studies tell us? *PLoS ONE* **2018**, *13*, e0205525. [CrossRef]
34. Behm, D.G.; Young, J.D.; Whitten, J.H.D.; Reid, J.C.; Quigley, P.J.; Low, J.; Li, Y.; Lima, C.D.; Hodgson, D.D.; Chaouachi, A.; et al. Effectiveness of traditional strength vs. power training on muscle strength, power and speed with youth: A systematic review and meta-analysis. *Front. Physiol.* **2017**, *8*, 423. [CrossRef] [PubMed]
35. Mirwald, R.L.; Baxter-Jones, A.D.G.; Bailey, D.A.; Beunen, G.P. An assessment of maturity from anthropometric measurements. *Med. Sci. Sport. Exerc.* **2002**, *34*, 689–694. [CrossRef]
36. Glatthorn, J.F.; Gouge, S.; Nussbaumer, S.; Stauffacher, S.; Impellizzeri, F.M.; Maffiuletti, N.A. Validity and reliability of Optojump photoelectric cells for estimating vertical jump height. *J. Strength Cond. Res.* **2011**, *25*, 556–560. [CrossRef] [PubMed]
37. Caffrey, E.; Docherty, C.L.; Schrader, J.; Klossner, J. The ability of 4 single-limb hopping tests to detect functional performance deficits in individuals with functional ankle instability. *J. Orthop. Sports Phys. Ther.* **2009**, *39*, 799–806. [CrossRef] [PubMed]
38. Le Naour, T.; Ré, C.; Bresciani, J.P. 3D feedback and observation for motor learning: Application to the roundoff movement in gymnastics. *Hum. Mov. Sci.* **2019**, *66*, 564–577. [CrossRef]
39. Cohen, J. A power primer. *Psychol. Bull.* **1992**, *112*, 155–159. [CrossRef]
40. Georgopoulos, N.A.; Markou, K.B.; Theodoropoulou, A.; Vagenakis, G.A.; Benardot, D.; Leglise, M.; Dimopoulos, J.C.A.; Vagenakis, A.G. Height velocity and skeletal maturation in elite female rhythmic gymnasts. *J. Clin. Endocrinol. Metab.* **2001**, *86*, 5159–5164. [CrossRef]
41. Bogdanis, G.C.; Donti, O.; Papia, A.; Donti, A.; Apostolidis, N.; Sands, W.A. Effect of Plyometric Training on Jumping, Sprinting and Change of Direction Speed in Child Female Athletes. *Sports* **2019**, *7*, 116. [CrossRef]
42. Hall, E.; Bishop, D.C.; Gee, T.I. Effect of plyometric training on handspring vault performance and functional power in youth female gymnasts. *PLoS ONE* **2016**, *11*, e0148790. [CrossRef]
43. Sands, W.A.; Shultz, B.B.; Newman, A.P. Women's gymnastics injuries: A 5-year study. *Am. J. Sports Med.* **1993**, *21*, 271–276. [CrossRef]

44. McAuliffe, S.; Tabuena, A.; McCreesh, K.; O'Keeffe, M.; Hurley, J.; Comyns, T.; Purtill, H.; O'Neill, S.; O'Sullivan, K. Altered strength profile in Achilles tendinopathy: A systematic review and meta-analysis. *J. Athl. Train.* **2019**, *54*, 889–900. [CrossRef] [PubMed]
45. Stevenson, H.; Beattie, C.S.; Schwartz, J.B.; Busconi, B.D. Assessing the Effectiveness of Neuromuscular Training Programs in Reducing the Incidence of Anterior Cruciate Ligament Injuries in Female Athletes: A Systematic Review. *Am. J. Sports Med.* **2015**, *43*, 482–490. [CrossRef] [PubMed]
46. Bradshaw, E.J.; Hume, P.A. Biomechanical approaches to identify and quantify injury mechanisms and risk factors in women's artistic gymnastics. *Sports Biomech.* **2012**, *11*, 324–341. [CrossRef] [PubMed]
47. Mkaouer, B.; Jemni, M.; Amara, S.; Chaabène, H.; Tabka, Z. Kinematic and kinetic analysis of two gymnastics acrobatic series to performing the backward stretched somersault. *J. Hum. Kinet.* **2013**, *37*, 17–26. [CrossRef]
48. Nagahara, R.; Takai, Y.; Haramura, M.; Mizutani, M.; Matsuo, A.; Kanehisa, H.; Fukunaga, T. Age-Related Differences in Spatiotemporal Variables and Ground Reaction Forces During Sprinting in Boys. *Pediat. Exerc. Sci.* **2018**, *30*, 335–344. [CrossRef]

Article

Gastrocnemius Medialis Architectural Properties in Flexibility Trained and Not Trained Child Female Athletes: A Pilot Study

Ioli Panidi, Gregory C. Bogdanis *, Vasiliki Gaspari, Polyxeni Spiliopoulou, Anastasia Donti, Gerasimos Terzis and Olyvia Donti *

Sports Performance Laboratory, School of Physical Education & Sport Science, National and Kapodistrian University of Athens, 17237 Athens, Greece; ipanidi@phed.uoa.gr (I.P.); vgaspari@phed.uoa.gr (V.G.); spipolyxeni@phed.uoa.gr (P.S.); adonti@phed.uoa.gr (A.D.); gterzis@phed.uoa.gr (G.T.)

* Correspondence: gbogdanis@phed.uoa.gr (G.C.B.); odonti@phed.uoa.gr (O.D.); Tel.: +30-21-07276115 (G.C.B.); +30-21-07276109 (O.D.)

Received: 15 January 2020; Accepted: 2 March 2020; Published: 4 March 2020

Abstract: Gastrocnemius medialis (GM) architecture and ankle angle were compared between flexibility trained (n = 10) and not trained (n = 6) female athletes, aged 8–10 years. Ankle angle, fascicle length, pennation angle and muscle thickness were measured at the mid-belly and the distal part of GM, at rest and at the end of one min of static stretching. Flexibility trained (FT) and not trained athletes (FNT) had similar fascicle length at the medial (4.19 ± 0.37 vs. 4.24 ± 0.54 cm, respectively, $p = 0.841$) and the distal part of GM (4.25 ± 0.35 vs. 4.18 ± 0.65 cm, respectively, $p = 0.780$), similar pennation angles, and muscle thickness ($p > 0.216$), and larger ankle angle at rest (120.9 ± 4.2 vs. 110.9 ± 5.8°, respectively, $p = 0.001$). During stretching, FT displayed greater fascicle elongation compared to FNT at the medial (+1.67 ± 0.37 vs. +1.28 ± 0.22 cm, respectively, $p = 0.048$) and the distal part (+1.84 ± 0.67 vs. +0.97 ± 0.97 cm, respectively, $p = 0.013$), larger change in joint angle and muscle tendon junction displacement (MTJ) ($p < 0.001$). Muscle thickness was similar in both groups ($p > 0.053$). Ankle dorsiflexion angle significantly correlated with fascicle elongation at the distal part of GM ($r = -0.638$, $p < 0.01$) and MTJ displacement ($r = -0.610$, $p < 0.05$). Collectively, FT had greater fascicle elongation at the medial and distal part of GM and greater MTJ displacement during stretching than FNT of similar age.

Keywords: youth; fascicle length; muscle thickness; maturation; stretching exercises; musculotendinous junction; ultrasound

1. Introduction

Triceps-surae (gastrocnemius medialis, gastrocnemius lateralis, and soleus) muscles architecture is an important functional characteristic in athletes, patients, and the elderly [1,2], as these muscles are prime movers of the ankle joint during locomotion [3,4]. Longer and less pennate gastrocnemius muscle fascicles increase muscle shortening velocity and thus power output [5] while pennation angle and muscle thickness are positively correlated with muscle force production [6,7].

Maturational growth, from infant to adult, and mechanical stimuli alter triceps-surae muscle architecture [8,9]. During growth, muscles are continuously stretched due to skeletal development [10], but data on gastrocnemius architectural properties in developing children are limited [8,10,11]. Moreover, it is largely unknown how gastrocnemius muscles architecture is related to functional properties in youth athletes [8] although athletes participating in sports like gymnastics, figure skating or ballet, are submitted to regular flexibility training [12].

Various training modes, such as resistance and eccentric training, provide the mechanical stimulus to induce morphological changes in the muscle-tendon unit by altering fascicle length, muscle thickness and volume, and pennation angle [13,14]. The effect of these types of training on muscle structure is well documented however, evidence is limited on stretching interventions [15], although flexibility is considered a major component of physical fitness [16]. Long-term static stretching interventions in humans, examining differences in muscle architecture and joint range of motion (ROM), presented equivocal results [17,18]. For example, Freitas and Mil-Homens [17] found a significant increase in biceps femoris fascicle length (+12.3 mm, $p = 0.04$) in physically active participants, following 8 weeks of intensive static stretching training (450 sec of stretching repeated 3 times per week). In contrast, Lima et al. [18] did not observe any changes in biceps femoris and vastus lateralis muscles architecture after 8 weeks of training in 12 physically active participants, using a short-duration, static stretching (3 sets of 30 sec of stretching, 3 times per week). The discrepant results between studies may be due to the different stretching training volume and intensity, combined with the short duration of the interventions (~8 weeks) [15]. Along this line, recent cross-sectional studies that examined populations with a chronic flexibility training background (>15 years of systematic stretching) reported that professional ballet dancers [11] and elite level rhythmic gymnasts [19] had longer resting fascicle length in gastrocnemius medialis compared to controls or not trained in flexibility athletes. These studies highlight that muscle architecture differs between athletes with different flexibility-training history; although sport-specific selection criteria or heredity may also be reflected in the dissimilarities in muscle architecture observed. Therefore, examining differences in muscle architecture in youth athletes submitted to different training load characteristics, provides useful information on muscle longitudinal growth in typically developing children and allows for the definition of exercise prescription in clinical populations. To this end, this study examined differences in gastrocnemius medialis (GM) architectural properties at rest and during stretching between child rhythmic gymnasts who trained and competed for at least 2 years, with same age girls participating in volleyball training. Rhythmic gymnasts were selected for their large joint ROM and compliant muscles and also due to their extensive flexibility training [20] while volleyball players were selected because their training included much less stretching training volume [21]. It was hypothesized that flexibility trained child female athletes would have longer fascicles at rest and during stretching compared to same age, flexibility not trained athletes.

2. Materials and Methods

2.1. Subjects

Ten female rhythmic gymnasts and six volleyball players, aged 8–10 years, took part in this study. Rhythmic gymnasts who competed at Hellenic, age-group (8–10 years old) all-around competitions were recruited from different gymnastics clubs and represented the flexibility trained group (FT). Volleyball players were participating in volleyball training in one club and were considered as not trained in flexibility athletes (FNT). Both sports, rhythmic gymnastics and volleyball, involve weight-bearing activities, but, rhythmic gymnastics training includes systematic stretching (≈40–60 min per session), while volleyball training includes <10 min of stretching exercises per session [22–25]. Athletes' characteristics are shown in Table 1. Maturity offset was calculated according to Mirwald et al. [26]. Before participating in the study, the athletes and their parents were informed about the aim and procedures of the study and provided written informed consent. The athletes had no injuries of the lower limbs for the past six months. The study design and procedures were in accordance with the declaration of Helsinki. The Institutional Ethics Committee approved the study (registration number: 1040, 14 February 2018).

Table 1. Anthropometric characteristics of the participants (means ± standard deviation).

Anthropometric Characteristics	Flexibility Trained Athletes (n = 10)	Flexibility Untrained Athletes (n = 6)	t (14)	p
Age (y)	9.00 ± 0.56	9.00 ± 0.63	0.000	1.000
Training experience (y)	3.70 ± 1.25	2.67 ± 0.52	1.906	0.770
Height (m)	1.34 ± 0.61	1.38 ± 0.3	−1.388	0.187
Body mass (kg)	27.57 ± 3.44	40.15 ± 5.89	−5.450	0.000
Body Mass Index(kg/m^2)	15.29 ± 1.15	21.05 ± 2.39	−6.565	0.000
Knee height (cm)	29.20 ± 2.03	30.32 ± 2.08	−1.057	0.308
Leg length (cm)	69.30 ± 3.62	70.33 ± 2.34	−0.622	0.544
Maturity offset	−5.52 ± 0.17	−4.99 ± 0.26	−5.023	0.000

2.2. Experimental Design

In order to examine differences in gastrocnemius medialis (GM) architectural properties and ankle angle between child athletes with different flexibility training history, all participants were tested over two sessions. In the first, familiarization session, athletes became familiar with the study protocol. Anthropometric characteristics of the athletes were also assessed during this session. In the main testing session, athletes' ankle angle and GM architectural characteristics were assessed in two conditions: (a) at rest and (d) during stretching. Resting ankle joint angle and GM architecture (fascicle length, pennation angle, muscle thickness) were assessed with the athletes lying in prone position on a physiotherapy bed for 20 min. (for detailed information, see description below). Following measurements of ankle angle and GM architecture at rest, athletes were standing for two min. Then, athletes performed a 1-min standing ankle dorsiflexion. Five seconds before the end of the stretching intervention a pause was imposed to obtain still ultrasound images. At the end of the 1 min stretching maximal ankle dorsiflexion was also assessed (for detailed information, see description below). No intense exercise or stretching was allowed in the 48 h preceding testing.

Anthropometric Characteristics

Height was measured without shoes with the use of a stadiometer, and body mass was measured with a calibrated digital scale (Seca 208 and Seca 710, Hamburg, Germany). Body mass index was calculated as the ratio of body weight to the squared standing height (kg/m^2). The length of each lower extremity was measured as the distance between trochanter major to the floor with the participants in standing position. The distance between tibiofemoral joint cleft and medial malleolus was determined as calf length.

2.3. Gastrocnemius Medialis Architecture and Ankle Joint Angle at Rest

In order to avoid trigonometric estimations or multiple scans along the muscle length to be assembled [27], in the present study, panoramic ultrasound images were obtained, via extended-field-of view imaging, along the fascicle length of GM.

All ultrasound measurements were performed in the morning and after athletes remained in a prone position on the examination bed, with their ankles hanging loosely on the outside of the bed for at least 20 min [28]. Muscle architecture of the right leg GM, (dominant leg that is in stance while kicking a ball) was obtained with a 10 MHz linear probe (38 mm) via extended field of view mode (Product model Z5, Shenzhen, Mindray Bio-Medical Electronics Co., Ltd., Shenzhen, China). Ultrasound images were recorded at the medial and the distal part of the GM muscle belly: one-third and half of the distance from the popliteal crease to the center of the medial malleolus, respectively. These points were marked on the skin using an echo-absorptive tape that served as reference marker [28] (Figure 1). In order to measure musculotendinous junction (MTJ) displacement, MTJ was located by real-time static ultrasound imaging and marked on the skin by an echo-absorptive tape, as well. The transducer was orientated perpendicular to the skin and parallel to the fascicles to minimize perspective and parallax measurement errors [29]. A probe path (dashed line) was drawn on the skin with a permanent pen by using static ultrasound according to the fascicle path seen from the ultrasound image. A single view was taken by moving continuously the probe in a slow and steady rate along the marked path. For each part of the muscle (medial and distal), three different fascicle lengths were measured from the deep aponeurosis to the superficial aponeurosis with a linear trace. Where the muscle fascicles met the lower aponeurosis the respective pennation angles were measured. The average of the lengths and angles of the three fascicles was used for statistical analysis for each part of the GM. The distance between the superficial and deep aponeuroses was determined as muscle thickness. Two consecutive measurements for each part of the muscle were assessed, and the average value was used for further analysis. All images were analyzed with image analysis software (Motic Images Plus 2.0, Motic, Hong Kong, China). Test-retest reliability was determined by using the intraclass correlation coefficient on 6 participants, on two separate days. The ICC (two-way random effects) for muscle fascicle length was 0.93 (95% CI: 0.576–0.990, $p = 0.000$), for muscular thickness it was 0.90 (95% CI: 0.474–0.984, $p = 0.001$), and for pennation angle, 0.95 (95% CI: 0.689–0.993, $p = 0.001$).

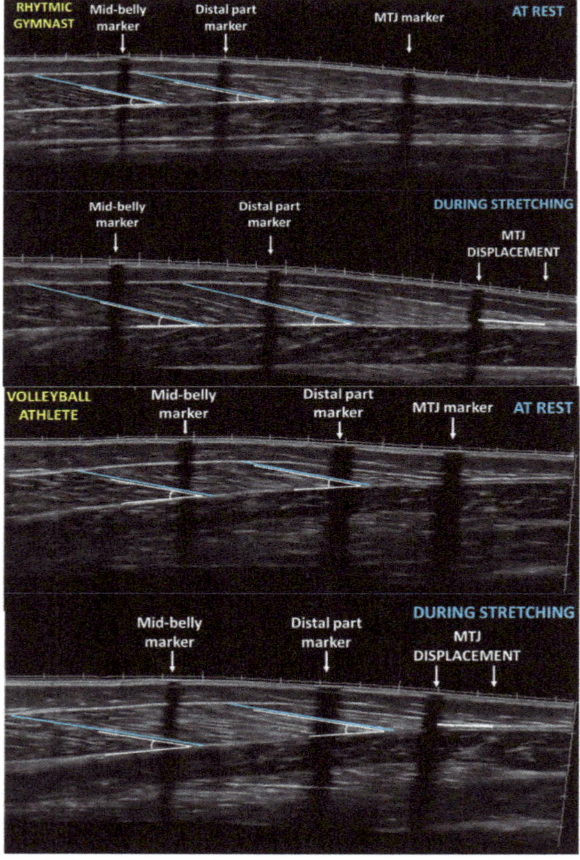

Figure 1. Panoramic sonographic image of gastrocnemius medialis of a rhythmic gymnast (top panel) and a volleyball player (bottom panel) at rest and during stretching showing fascicle length and pennation angle at the mid-belly and at the distal part of GM. MTJ: muscle-tendon junction.

Ankle joint angle at rest was also measured with the athletes lying in a prone position, with their ankles hanging loosely off the bed. Resting ankle joint angle was defined as the angle created by the intersection of the femur-tibia to lateral malleolus line and lateral malleolus to fifth metatarsal line [19]. Reflective markers were placed on these anatomical points in order to define the ankle angle using a digital camera (Casio Exilim Pro EX-F1, Shibuya, Tokyo, Japan). Image analysis was performed via a free software (Tracker 4.91© 2016 Douglas Brown). Intra-class correlation coefficients for resting ankle angle was 0.98 (95% Confidence Intervals (CI): 0.833–0.998, $p = 0.000$).

2.4. Gastrocnemius Medialis Architecture and Ankle Joint Angle during Ankle Dorsiflexion Stretching

Panoramic ultrasound images from the two parts of the GM muscle belly (medial and distal) were obtained following the method described above. Reflective motion analysis markers, and echo-absorptive tapes remained on the skin, and the drawn path (dashed line) of the resting measurements was used, to re-assess the regions of interest. Following two minutes of standing, all athletes performed a slow, passive standing dorsiflexion stretching, for one minute. Five seconds before the end of the stretching intervention a pause was imposed to capture still images. To obtain GM ultrasound images during stretching, the probe was placed 38 mm above the skin marker that identified the middle part of the muscle belly. In addition, 'MTJ displacement' was defined as the difference between the MTJ marker at rest and MTJ point during stretching (Figure 1).

Ankle dorsiflexion stretching while standing is commonly performed in sport practice [30], and the athletes were familiar with it. Stretching was performed with the athletes barefoot. Athletes were instructed to relax while they passively stretched their ankle plantar flexors, in a slow and continuous manner. The foot to be tested (right) was placed on the midline of a marked area on the floor, and the left foot was placed forward at step-length distance. The end point of standing dorsiflexion stretching was defined as the point that the athletes felt discomfort without lifting their heel and with no pelvic rotation. The athletes put their hands against the wall to maintain balance and were asked to keep the extended position of their hip and knee joints, during stretching [30]. Stretch intensity was indicated by the athletes using the 0–10 Wong-Baker FACES Pain Scale for children [31] to ensure that stretch achieved the point of discomfort (~8 on a scale of 0–10). During the execution of the stretch, participants were instructed to reach a pain of discomfort level of 8 in the scale of 0–10, and thus they held the stretch at exactly this perceived intensity. The psychometric properties of this commonly used pictorial scale assessing acute pain have been found to be appropriate for children over the age of 3 [31]. Six faces depict different expressions, ranging from "no hurt" to "extremely upset from pain". A digital camera (Casio Exilim Pro EX-F1, Shibuya, Tokyo, Japan) was placed perpendicular to the plane of motion of the right leg in order to record the standing ankle dorsiflexion angle. Stretching ankle joint angle was analyzed using reflective markers placed on the knee, ankle and fifth metatarsal and calculated using free software (Tracker 4.91© 2016 Douglas Brown). Maximal standing dorsiflexion was defined as the intersection of a line joining the knee and ankle markers and horizontal (a line crossing the heel and the fifth metatarsal).

2.5. Statistical Procedures

Descriptive statistics were calculated. Shapiro-Wilks test checked for normality of data distribution. Pearson correlations coefficient (r) detected linear relations between the examined variables. Unpaired T-test examined differences between groups in anthropometry and architectural characteristics of GM at rest. A two-way ANOVA (time x group) with repeated measures for time (rest or stretch) and group (flexibility trained vs. not trained) was conducted separately for the medial and the distal part of the muscle, to examine the effect of stretching on fascicle lengths, pennation angles and thicknesses, and ankle joint angle. A Tukey post-hoc test was performed when a significant main effect or interaction was observed ($p < 0.05$). Effect sizes calculation for pairwise comparisons was performed with Cohen's d [32]. To assess test-retest reliability the intra-class correlation coefficients (ICCs) were used. Statistical significance was set at $p < 0.05$. Statistical analyzes were conducted using SPSS (SPSS Statistics Version 25.0, IBM corporation, Armonk, NY, USA).

3. Results

3.1. Gastrocnemius Medialis Architecture and Ankle Joint Angle at Rest

Resting fascicle length of FT athletes was similar to FNT at the mid-belly (4.19 ± 0.37 vs. 4.24 ± 0.54 cm, respectively, $t_{14} = -0.204$, $p = 0.841$) and the distal part of gastrocnemius medialis (4.25 ± 0.35 vs. 4.18 ± 0.65 cm, respectively, $t_{14} = 0.284$, $p = 0.780$). FT and FNT athletes displayed also similar pennation angle and muscle thickness at the mid-belly ($t_{14} = 0.661$, $p = 0.519$ and $t_{14} = 0.002$, $p = 0.998$, respectively) and the distal part of the gastrocnemius medialis ($t_{14} = -1.297$, $p = 0.216$ and $t_{14} = 0.807$, $p = 0.433$, respectively) (Table 2). Resting ankle angle was larger in FT by 8% compared with FNT (120.86 ± 4.19° vs. 110.95 ± 5.79°, respectively, $t_{14} = 3.982$, $p = 0.001$) (Table 2).

Table 2. Changes in muscle architecture characteristics and ankle angle following stretching for the flexibility trained (FT) (n = 10) and not trained athletes (FNT) (n = 6).

Variables	Athletes	Pre–Stretching Measurements	Δ Values (Pre–vs. Stretching)	*p*	Cohens' *d* (Pre–Post Stretching)	Cohens' *d* of Δ Values between Groups
Fascicle Length Mid–belly (cm)	FT FNT	4.19 ± 0.37 4.24 ± 0.54	+1.67 ± 0.39 +1.28 ± 0.24	0.048	5.30 2.60	1.21
Fascicle Length Distal part (cm)	FT FNT	4.25 ± 0.35 4.18 ± 0.65	+1.84 ± 0.70 +0.97 ± 0.32	0.013	4.58 1.62	1.59
Thickness Mid–belly (cm)	FT FNT	1.53 ± 0.12 1.53 ± 0.24	+0.23 ± 0.06 +0.15 ± 0.10	0.053	1.94 0.73	1.25
Thickness Distal part (cm)	FT FNT	0.74 ± 0.22 0.64 ± 0.25	+0.46 ± 0.16 +0.29 ± 0.16	0.061	2.31 1.20	1.14
Pennation angle Mid–belly (°)	FT FNT	21.76 ± 1.76 21.19 ± 1.43	−4.93 ± 2.01 −2.90 ± 1.41	0.048	3.57 2.09	1.19
Pennation angle Distal part (°)	FT FNT	18.00 ± 1.96 19.46 ± 2.51	−2.48 ± 2.60 −1.35 ± 1.70	0.362	1.44 0.69	0.52
Ankle angle (°)	FT FNT	120.9 ± 4.2 111.0 ± 5.8†	−62.7 ± 6.7 −44.9 ± 6.3	0.001	13.25 8.92	2.88
MTJ Displacement (cm)	FT FNT		+2.31 ± 0.40 +1.54 ± 0.30 †	0.001		2.24

†: $p < 0.001$ from the corresponding value in flexibility trained athletes.

3.2. Gastrocnemius Medialis Architecture and Ankle Joint Angle during Ankle Dorsiflexion Stretching

During stretching, the elongation of fascicles was greater in FT athletes compared to the FNT at the mid-belly of the muscle by 23% (+1.67 ± 0.37 cm vs. +1.28 ± 0.22 cm, $p = 0.048$) and the distal part by 47% (+1.84 ± 0.67 vs. +0.97 ± 0.29 cm, $p = 0.013$). Furthermore, FT athletes displayed greater maximal ankle dorsiflexion by 13% ($p < 0.001$), as well as greater muscle tendon junction displacement by 33% ($p < 0.001$) (Table 2). No differences were found between groups in muscle thickness at mid-belly and at the distal part ($p > 0.053$). However, FNT athletes displayed greater pennation angle at the mid-belly of gastrocnemius medialis (−2.90 ± 1.29 vs. −4.93 ± 1.91 vs., $p = 0.048$), but not at the distal part ($p = 0.362$) (Table 2).

3.3. Correlations Between Fascicle Length, Ankle Angles and MTJ Displacement

When all athletes were considered as a group, significant correlations were found between fascicle elongation at the distal part of GM and MTJ displacement ($r = 0.752$, $p < 0.01$) and ankle angle during stretching ($r = -0.638$, $p < 0.01$). Moreover, a significant correlation was found between MTJ displacement and ankle angle during stretching ($r = -0.610$, $p < 0.05$).

4. Discussion

This study examined differences in GM architectural properties at the middle and the distal part of the muscle belly, at rest and during stretching, between flexibility trained and not trained female athletes (rhythmic gymnasts and volleyball players, respectively), aged 8–10 years. The main finding of this study was that, at rest, the two groups displayed similar GM architectural properties, but during stretching FT displayed greater fascicle elongation at the middle and the distal part of GM, and greater MTJ displacement. In addition, FT had larger ankle joint angles at rest and larger change in ankle angle during stretching, compared with FNT, athletes. Significant correlations were found between fascicle elongation at the distal part of GM, MTJ displacement and ankle angle during dorsiflexion.

Gastrocnemius muscle is a prime mover in ankle plantar flexion and thus its architecture is related to force/power production and range of motion [5]. However, chronic modifications to gastrocnemius muscles architecture because of exercise or training in children are currently unknown [33–35]. The results of this study indicated that the two groups had similar resting fascicle length at the medial and the distal part of GM. This finding is interesting because recent cross-sectional studies found longer resting fascicle length in flexibility trained, compared with untrained adult participants. For example, a previous study [11] compared professional ballet dancers to controls and found that ballet dancers had longer fascicles in GM in resting prone position (55 ± 5 vs. 47 ± 6 mm, respectively). Another study that examined elite rhythmic gymnasts and female volleyball players, also reported that gymnasts had longer fascicle length at rest at the mid-belly and the distal part of GM compared to volleyball players, by 20 and 18%, respectively [19]. Resting fascicle length has been recently linked with plantar flexion torque and work in healthy adults [34] and was related to the muscle's force-length relationship [35]. The participants of the present study were growing children, aged 8–10 years. During growth, muscle-tendon units are increased in length, to keep up with increases in bone length [10,35]. Benard et al. [10] examined how maturational growth and skeletal development imparts changes in muscle architecture and found that GM muscle length increases (through an increase in muscle, tendon and fascicle length) approximately 6% per year from age 5 to 12, in proportion with increases in tibia length. That study also reported that the length component of the physiological cross-sectional area of GM as well as muscle fascicles, increased in length [10]. Thus, even if long-term systematic and extensive flexibility training might increase muscle fascicle length, it is plausible that the mechanical stimulus of stretching training is not adequate to induce changes additional to maturational growth in developing children. Nevertheless, cross-sectional study-designs do not imply causation and, at present, the chronic effect of static stretching training on joint range of motion and muscle architecture in humans is not sufficiently documented.

In the present study, fascicle elongation was measured at mid-belly and at the distal part of GM during maximal ankle dorsiflexion. This was because the mid-belly might not accurately reflect muscle architecture across the entire gastrocnemius muscle [36]. The results of this study indicated that fascicle elongation was greater in FT athletes compared to FNT at the middle ($p = 0.048$, $d = 1.21$) and the distal part of GM, ($p = 0.013$, $d = 1.59$) by 23 and 47%, respectively. This result is in line with previous research reporting greater elongation of GM fascicles in flexible compared to inflexible subjects [19,37]. Importantly, an almost twofold greater elongation was observed in FT athletes at the distal part of GM compared with FNT (+1.84 ± 0.70 vs. 0.97 ± 0.32 cm, respectively, $p = 0.013$) and significantly greater MTJ displacement ($p = 0.001$, $d = 2.24$). Simpson et al. [38] examined adaptations in architectural characteristics of gastrocnemius medialis and lateralis following 6 weeks of stretching training, in adult participants. The authors reported that muscle fascicles in the belly increased by 5.1% by week 6 whereas fascicles in the junction were 25% longer [38]. A previous study also reported that adult rhythmic gymnasts displayed greater fascicle elongation at the distal part compared to volleyball players (45 vs. 39%, respectively, $p = 0.026$) [19]. This finding highlights that there may be non-uniform morphological adaptations along the length of a bi-articular muscle, like GM, depending on training history. Chronic flexibility training and/or other components of sport-specific training may induce muscle architectural adaptations that differ between the muscle belly and the region near the musculotendinous junction. Previous animal studies identified higher levels of myosin heavy chain mRNA at the MTJ of fibers stretched for 4 days [39] and suggested that fiber lengthening, following stretching, created a need for contractile protein synthesis and assembly into myofibrils at the MTJ [40]. A recent study in humans also reported that following a 4-weeks resistance training intervention, the remodeling of muscle fibres near the MTJ was very high [41].

Enhanced joint ROM following static stretching training has been shown with various stretching protocols in youth athletes or in physical education settings [42,43]. This study examined ankle angle at rest lying in prone position, and during maximal ankle dorsiflexion. The results of the present study indicated that at rest, rhythmic gymnasts had greater ankle angle by 8%, compared with volleyball players ($p = 0.001$, $d = 1.21$) (Table 2). A similar finding was reported in a previous study with adult rhythmic gymnasts, and the authors assumed that different resting ankle joint angle between groups may imply a different slack length in the muscles surrounding the ankle joint due to long-term, extensive flexibility training [19]. It is not known whether flexibility training and/or other components of sport-specific training may alter the "neutral", resting ankle joint angle [11]. Previous studies in adults reported similar ankle joint angles at rest between flexibility trained and not trained subjects [11,37]; however, further research is required on the impact of chronic flexibility training on body tissues determining joints range of motions.

Ankle joint dorsiflexion angle was also significantly greater in FT compared to FNT athletes by 13% ($p = 0.001$, $d = 2.88$), and muscle tendon junction displacement by 33% ($p < 0.001$, $d = 2.24$) (Table 2). Moltubakk et al. [11] and Donti et al. [19] also found larger ankle dorsiflexion angle in adult ballet dancers and gymnasts compared to controls, a fact mirroring their regular, intensive stretching training. Acute increases in joint ROM following stretching are mainly due to an increased tolerance to stretch [44]. The association of chronic increases in joint ROM with adaptations in muscle architecture has not been clearly established [17,18,37]. Some previous long-term stretching interventions in adults, indicated enhanced joint ROM followed by concomitant increases in fascicle length [15,38] while other long-term stretching interventions failed to detect changes in muscle architecture [18,37]. Amongst the factors determining joint ROM, maximal fascicle elongation at the distal part of the muscle belly and MTJ displacement in the present study, were strongly associated with larger maximal ankle dorsiflexion angle ($r = -0.638$, $p < 0.01$, and $r = -0.610$, $p = 0.05$, respectively). However, the cross-sectional design of this study limits interpretation of these findings. In addition, available studies indicate that there is considerable variation in GM muscle architecture associated with chronological age [8]. Thus, the small number of participants in this study is a limitation that should be acknowledged. Chronic intervention studies are required in developing athletes, to distinguish genetic or acquired through years of sport-specific training changes in muscle architecture in order to examine the contribution of changes in fascicle length to the increase in muscle length in typically developing children. It should be noted that the time frame of middle childhood (6–11 years) has been proposed as a 'window of opportunity' for developing flexibility and as a sensitive period for morphological changes [45,46].

5. Conclusions

Collectively, greater muscle elongation at the mid-belly and the distal part of GM during static stretching, and greater ankle angles at rest and during dorsiflexion were observed in FT compared to FNT female athletes, aged 8–10 years. These findings indicate that between children with different flexibility training history, muscle architecture differs only during stretching, and that there are non-uniform adaptations along GM length depending on training history. Albeit speculative, increased muscle fascicle elongation may represent an early-stage adaptation to stretching-induced increases in resting fascicle length found in flexibility trained, female adult athletes compared with not trained controls.

Author Contributions: Conceptualization, O.D., I.P., and G.C.B.; methodology, I.P., O.D., G.C.B., P.S., and G.T.; data curation, I.P., O.D., V.G.; writing—original draft preparation, O.D., I.P., A.D., writing—review and editing, O.D., G.C.B., and G.T.; supervision, O.D. All authors have read and agreed to the published version of the manuscript.

Funding: The authors received no external funding.

Acknowledgments: The authors would like to thank the athletes for their dedicated time and collaboration.

Conflicts of Interest: The authors have no conflict of interest to declare

References

1. Nesser, T.W.; Latin, R.W.; Berg, K.; Prentice, E. Physiological determinants of 40-m sprint performance in young male athletes. *J. Strength Cond. Res.* **1996**, *10*, 263–267. [CrossRef]
2. Graf, A.; Judge, J.O.; Õunpuu, S.; Thelen, D.G. The effect of walking speed on lower-extremity joint powers among elderly adults who exhibit low physical performance. *Arch. Phys. Med. Rehabil.* **2005**, *86*, 2177–2183. [CrossRef]
3. Bobbert, M.F.; Mackay, M.; Schinkelshoek, D.; Huijing, P.A.; van Ingen Schenau, G.J. Biomechanical analysis of drop and countermovement jumps. *Eur. J. Appl. Physiol. Occup. Physiol.* **1986**, *54*, 566–573. [CrossRef] [PubMed]
4. Ellis, R.G.; Sumner, B.J.; Kram, R. Muscle contributions to propulsion and braking during walking and running: Insight from external force perturbations. *Gait Posture* **2014**, *40*, 594–599. [CrossRef] [PubMed]
5. Baxter, J.R.; Hullfish, T.J.; Chao, W. Functional deficits may be explained by plantarflexor remodeling following Achilles tendon rupture repair: Preliminary findings. *J. Biomech.* **2018**, *79*, 238–242. [CrossRef] [PubMed]
6. Randhawa, A.; Wakeling, J.M. Associations between muscle structure and contractile performance in seniors. *Clin. Biomech.* **2013**, *28*, 705–711. [CrossRef]
7. Morse, C.I.; Thom, J.M.; Reeves, N.D.; Birch, K.M.; Narici, M.V. In vivo physiological cross-sectional area and specific force are reduced in the gastrocnemius of elderly men. *J. Appl. Physiol.* **2005**, *99*, 1050–1055. [CrossRef]
8. Legerlotz, K.; Smith, H.K.; Hing, W.A. Variation and reliability of ultrasonographic quantification of the architecture of the medial gastrocnemius muscle in young children. *Clin. Physiol. Funct. Imaging* **2010**, *30*, 198–205. [CrossRef]
9. Lichtwark, G.A.; Wilson, A.M. Optimal muscle fascicle length and tendon stiffness for maximising gastrocnemius efficiency during human walking and running. *J. Theor. Biol.* **2008**, *252*, 662–673. [CrossRef]
10. Benard, M.R.; Harlaar, J.; Becher, J.G.; Huijing, P.A.; Jaspers, R.T. Effects of growth on geometry of gastrocnemius muscle in children: A three-dimensional ultrasound analysis. *J. Anat.* **2011**, *219*, 388–402. [CrossRef]
11. Moltubakk, M.M.; Magulas, M.M.; Villars, F.O.; Seynnes, O.R.; Bojsen-Møller, J. Specialized properties of the triceps surae muscle-tendon unit in professional ballet dancers. *Scand. J. Med. Sci. Sports* **2018**, *28*, 2023–2034. [CrossRef] [PubMed]
12. Sands, W.A.; McNeal, J.R.; Panitente, G.; Murray, S.R.; Nassar, L.; Jemni, M.; Mizuguchi, S.; Stone, M.H. Stretching the spines of gymnasts: A review. *Sports Med.* **2016**, *46*, 315–327. [CrossRef] [PubMed]
13. Blazevich, A.J. Effects of physical training and detraining, immobilisation, growth and aging on human fascicle geometry. *Sports Med.* **2006**, *36*, 1003–1017. [CrossRef] [PubMed]
14. Franchi, M.V.; Reeves, N.D.; Narici, M.V. Skeletal muscle remodeling in response to eccentric vs. concentric loading: Morphological, molecular, and metabolic adaptations. *Front. Physiol.* **2017**, *8*, 447. [CrossRef]
15. Freitas, S.R.; Mendes, B.; Le Sant, G.; Andrade, R.J.; Nordez, A.; Milanovic, Z. Can chronic stretching change the muscle-tendon mechanical properties? A review. *Scand. J. Med. Sci. Sport.* **2018**, *28*, 794–806. [CrossRef]
16. Nuzzo, J.L. The Case for Retiring Flexibility as a Major Component of Physical Fitness. *Sports Med.* **2019**, 1–18. [CrossRef]
17. Freitas, S.R.; Mil-Homens, P. Effect of 8-week high-intensity stretching training on biceps femoris architecture. *Scand. J. Med. Sci. Sports* **2015**, *29*, 1737–1740. [CrossRef]
18. E Lima, K.M.; Carneiro, S.P.; Alves, D.D.S.; Peixinho, C.C.; de Oliveira, L.F. Assessment of muscle architecture of the biceps femoris and vastus lateralis by ultrasound after a chronic stretching program. *J. Sport Med.* **2015**, *25*, 55–60. [CrossRef]
19. Donti, O.; Panidis, I.; Terzis, G.; Bogdanis, G.C. Gastrocnemius Medialis Architectural Properties at Rest and During Stretching in Female Athletes with Different Flexibility Training Background. *Sports* **2019**, *7*, 39. [CrossRef]

20. Donti, O.; Tsolakis, C.; Bogdanis, G.C. Effects of baseline levels of flexibility and vertical jump ability on performance following different volumes of static stretching and potentiating exercises in elite gymnasts. *J. Sports Sci. Med. Sports* **2014**, *13*, 105.
21. Malliaras, P.; Cook, J.L.; Kent, P. Reduced ankle dorsiflexion range may increase the risk of patellar tendon injury among volleyball players. *J. Sci. Med. Sport.* **2006**, *9*, 304–309. [CrossRef] [PubMed]
22. Moltubakk, M.M.H. Effects of Long-Term Stretching Training on Muscle-Tendon Morphology, Mechanics and Function. Ph.D. Thesis, Norges Idrettshøgskole, Oslo, Norway, 2019.
23. Douda, H.T.; Toubekis, A.G.; Avloniti, A.A.; Tokmakidis, S.P. Physiological and anthropometric determinants of rhythmic gymnastics performance. *Int. J. Sports Physiol. Perform.* **2008**, *3*, 41–54. [CrossRef] [PubMed]
24. Simenz, C.J.; Dugan, C.A.; Ebben, W.P. Strength and conditioning practices of National Basketball Association strength and conditioning coaches. *J. Strength Cond. Res.* **2005**, *19*, 495–504. [CrossRef] [PubMed]
25. Gabbett, T.; Georgieff, B. Physiological and anthropometric characteristics of Australian junior national, state and novice volleyball players. *J. Strength Cond. Res.* **2007**, *21*, 902–908. [CrossRef]
26. Mirwald, R.L.; Baxter-Jones, A.D.; Bailey, D.A.; Beunen, G.P. An assessment of maturity from anthropometric measurements. *Med. Sci. Sports Exerc.* **2002**, *34*, 689–694. [CrossRef]
27. Kawakami, Y.; Ichinose, Y.; Fukunaga, T. Architectural and functional features of human triceps surae muscles during contraction. *J. Appl. Physiol.* **1998**, *85*, 398–404. [CrossRef]
28. Reeves, N.D.; Maganaris, C.N.; Narici, M.V. Ultrasonographic assessment of human skeletal muscle size. *Eur. J. Appl. Physiol.* **2004**, *91*, 116–118. [CrossRef]
29. Noorkoiv, M.; Stavnsbo, A.; Aagaard, P.; Blazevich, A.J. In vivo assessment of muscle fascicle length by extended field-of-view ultrasonography. *J. Physiol.* **2010**, *109*, 1974–1979. [CrossRef]
30. Jung, D.Y.; Koh, E.K.; Kwon, O.Y.; Yi, C.H.; Oh, J.S.; Weon, J.H. Effect of medial arch support on displacement of the myotendinous junction of the gastrocnemius during standing wall stretching. *J. Orthop. Sports Phys. Ther.* **2009**, *39*, 867–874. [CrossRef]
31. Wong, D.L.; Hockenberry-Eaton, M.; Wilson, D.; Winkelstein, M.L.; Schwartz, P. *Wong's Essentials of Pediatric Nursing*, 6th ed.; Mosby: St. Louis, MO, USA, 2001; p. 1301.
32. Cohen, J. A power primer. *Psychol. Bull.* **1992**, *112*, 155–159. [CrossRef]
33. Lieber, R.L.; Friden, J. Functional and clinical significance of skeletal muscle architecture. *Muscle Nerve* **2000**, *23*, 1647–1666. [CrossRef]
34. Ema, R.; Akagi, R.; Wakahara, T.; Kawakami, Y. Training-induced changes in architecture of human skeletal muscles: Current evidence and unresolved issues. *J. Sports Med. Phys. Fit.* **2016**, *5*, 37–46. [CrossRef]
35. O'Brien, T.D.; Reeves, N.D.; Baltzopoulos, V.; Jones, D.A.; Maganaris, C.N. Muscle–tendon structure and dimensions in adults and children. *J. Anat.* **2010**, *216*, 631–642. [CrossRef] [PubMed]
36. Zhang, T.K.; Fortuna, R.; Herzog, W. Distal and proximal fascicle length changes in active and passive human gastrocnemius muscle. *J. Undergrad. Res.* **2015**, *5*.
37. Blazevich, A.J.; Cannavan, D.; Waugh, C.M.; Fath, F.; Miller, S.C.; Kay, A.D. Neuromuscular factors influencing the maximum stretch limit of the human plantar flexors. *J. Appl. Physiol.* **2012**, *113*, 1446–1455. [CrossRef]
38. Simpson, C.L.; Kim, B.D.H.; Bourcet, M.R.; Jones, G.R.; Jakobi, J.M. Stretch training induces unequal adaptation in muscle fascicles and thickness in medial and lateral gastrocnemii. *Scand. J. Med. Sci. Sports* **2017**, *27*, 1597–1604. [CrossRef]
39. Dix, D.J.; Eisenburg, B.R. Myosin mRNA accumulation and myofibrillar genesis at the myotendinous junction of stretched muscle fibres. *J. Cell Biol.* **1990**, *111*, 1885–1894. [CrossRef]
40. Antin, P.B.; Tokunaka, S.; Nachmias, V.T.; Holtzer, H. Role of stress fiber-like structures in assembling nascent myofibrils in myosheets recovering from exposure to ethyl methanesulfonate. *J. Cell Biol.* **1986**, *102*, 1464–1479. [CrossRef]
41. Jakobsen, J.R.; Jakobsen, N.R.; Mackey, A.L.; Koch, M.; Kjaer, M.; Krogsgaard, M.R. Remodeling of muscle fibers approaching the human myotendinous junction. *Scand. J. Med. Sci. Sports* **2018**, *28*, 1859–1865. [CrossRef]
42. Donti, O.; Papia, K.; Toubekis, A.; Donti, A.; Sands, W.A.; Bogdanis, G.C. Flexibility training in preadolescent female athletes: Acute and long-term effects of intermittent and continuous static stretching. *J. Sports Sci.* **2018**, *36*, 1453–1460. [CrossRef]

43. Mayorga Vega, D.; Merino-Marban, R.; Redondo-Martín, F.J.; Viciana, J. Effect of a one-session-per-week physical education-based stretching program on hamstring extensibility in schoolchildren. *Kinesiol. Int. J. Fundam. Appl.* **2017**, *49*, 101–108. [CrossRef]
44. Weppler, C.H.; Magnusson, S.P. Increasing muscle extensibility: A matter of increasing length or modifying sensation? *Phys. Ther.* **2010**, *90*, 438–449. [CrossRef] [PubMed]
45. Lloyd, R.S.; Oliver, J.L. The youth physical development model: A novel approach to long-term athletic development. *Strength Cond. J.* **2012**, *34*, 61–72. [CrossRef]
46. Malina, R.M.; Bouchard, C.; Bar-Or, O. *Growth, Maturation, and Physical Activity*, 2nd ed.; Human Kinetics: Champaign, IL, USA, 2004; pp. 215–220.

Article

Backward Running: Acute Effects on Sprint Performance in Preadolescent Boys

Dimitrios Petrakis [1,†], Eleni Bassa [1,*,†], Anastasia Papavasileiou [1], Anthi Xenofondos [2] and Dimitrios A. Patikas [1]

1 Laboratory of Evaluation of Human Biological Performance, Faculty of Physical Education and Sport Science, Aristotle University of Thessaloniki, 54124 Thessaloniki, Greece

2 Division of Kinesiology, School of Public Health, The University of Hong Kong, Pok Fu Lam 00000, Hong Kong

* Correspondence: lbassa@phed.auth.gr

† These authors contributed equally to this work.

Received: 19 February 2020; Accepted: 20 April 2020; Published: 23 April 2020

Abstract: The aim of this study was to examine the acute effect of backward running (BwR) during warm-up on a 20-m sprint of boys' performance, compared to forward running (FwR). Fourteen recreationally active preadolescent boys (aged 12.5 ± 0.5 years) were examined in 3 protocols: warm-up (control condition), warm-up with 3 × 10 m additional BwR sprints and warm-up with 3 × 10 m additional FwR sprints. Participants were evaluated 4 minutes after each protocol on a 20-m sprint and intermediate distances, as well as the rate of perceived exertion (RPE). Sprint speed across 10–20 m was significantly higher for the BwR warm-up compared to the regular warm-up ($p < 0.05$) and a significantly higher RPE after the BwR and FwR protocols compared to the control condition was recorded ($p < 0.05$). No significant difference was detected across the distances 0–5, 5–10, 0–10 and 0–20 m. Although adding 3 × 10-m sprints of BwR or FwR after the warm-up did not enhance performance in a 20 m sprint of preadolescent boys, the positive effect of BwR across 10–20 m distance suggests that BwR could be an alternative means for enhancing performance for certain phases of a sprint for this age. However, preadolescent boys' response to different sprint conditioning exercise stimuli and the optimization of rest time to maximize performance remain to be determined.

Keywords: preadolescence; child; post-activation performance enhancement; sprint; warm-up; rate of perceived exertion

1. Introduction

Warm-up, as a common practice applied prior to exercise and sports activities, has the potential to improve performance [1]. There are several mechanisms that may contribute to this, such as increased muscle temperature [2,3], the elevation of oxygen uptake kinetics [4] and changes in the function of the neuromuscular system [5]. The inclusion of conditioning exercises in a warm-up, i.e., high-intensity exercises, is widely thought to potentiate performance [6,7]. "Post-activation performance enhancement" (PAPE) is a new term introduced by Cuenca-Fernández et al. [8] and describes such effects. In contrast to the classic post-activation potentiation, i.e., an increase in twitch force and power after electrically or voluntarily induced intense contraction [9,10], PAPE has a longer and weaker effect on performance, and is more likely attributed to different mechanisms [11]; the former is attributed to the phosphorylation of the myosin regulatory light-chain and the latter to changes in muscle temperature, muscle/cellular water content and/or muscle activation [11]. However, the acute effects of different conditioning stimuli—especially during warm-up—on the performance of tasks such as sprinting, is yet to be determined.

The effects of conditioning stimuli on sprint performance have been previously tested in adults. Effective PAPE effects have been reported using different types of conditioning stimulus, such as high resistance loads [6,12,13] or jumping exercises [14]. Although it was previously emphasized that PAPE stimulus is more effective when it is biomechanically similar to the subsequent activity [15], studies using sprints as conditioning stimuli to enhance sprint performance are limited [16,17]. These studies showed that adults did not improve their 60-m sprint speed after 2 × 60 m sprints [16], whereas young male track and field athletes increased their speed in a 100 m sprint after 2 × 20 m resisted sprints, and not after the same sprints as conditioning without resistance [17]. It seems therefore that the properties of the conditioning stimulus might be critical for the outcome of the study.

Regarding young ages and development, PAPE has not been extensively investigated in children prior to puberty. Although there are no differences in post-activation potentiation of the plantar flexor muscles between men, adolescents and pre-adolescents [18], it has been shown that after maximal half-squats, PAPE in terms of squat jump height was apparent in adult men but not in women, or adolescents and children of both sexes [19]. Similarly, in preadolescent female gymnasts, high-intensity task-specific (Rondat) or non-specific medium-intensity (double tuck jumps) conditioning contractions were not adequate to induce PAPE on drop jumps [20]. Nonetheless, there are indications that young adolescents can benefit from conditioning stimuli in the long-term. More specifically, resistance exercise can cause PAPE effects in adolescents only after and not before 10 weeks of resistance and sprint training [21]. Hence, it seems that the open question is not whether children are capable of demonstrating PAPE, but which are the optimal conditions and the appropriate candidates of conditioning stimuli to achieve it.

Running backwards (BwR) or forwards (FwR) are common types of movement in several sports [22–24], but there are several functional differences between them. Compared to FwR, BwR demonstrates greater lower limb muscle activation [25], higher rate of force development [26] and lower mechanical stress on the knee [27]. These properties suggest that BwR could be a promising, safe and efficient training stimulus. Furthermore, FwR and BwR differ in the type of contractions involved during the task. More specifically, BwR is associated more with concentric and less with eccentric work on the lower limbs [28]. This issue is of particular importance, because there is evidence that children, are not efficient in tasks that incorporate eccentric contractions, such as vertical jumps [29,30] and FwR [31], since they demonstrate prolonged contact time with the ground and hence inadequate transfer of energy among the joints. On the other hand, BwR is an effective training method to improve in the long-term children's sprint speed [32], whilst there is no information regarding the acute effect of BwR on sprinting. To our knowledge, it is still unknown whether PAPE in pre-adolescent children's sprinting performance could be induced by implementing BwR in a warm-up, i.e., a stimulus with a greater concentric contraction profile than FwR. Therefore, it remains to be tested if this effect of BwR would be greater than a warm-up protocol including FwR, which relies more on eccentric contractions. Considering the above, the aim of this study was to examine the acute effect of 3 × 10 m BwR bouts compared to FwR during a warm-up, on sprint performance, in pre-adolescent boys. We hypothesized that BwR would potentiate performance in a 20 m sprint and intermittent distances more than a typical warm-up program or a typical warm-up with FwR. This information could be useful for seeking methods to optimize sprint performance in children after their warm-up.

2. Materials and Methods

2.1. Participants

Fourteen (n = 14) recreationally active preadolescent boys (age: 12.5 ± 0.5 years; body mass: 50.2 ± 10.5 kg; height: 159.4 ± 10.1 cm) volunteered to participate in the study. This sample size for the present experimental design corresponds to 0.8 power, for 0.65 effect size at a = 0.05 (G-power, v.3.1.9.4, University of Kiel, Kiel, Germany). Maturity offset from peak height velocity was calculated according to the prediction equation based on anthropometric measures, sex and age [33] and the participants were characterized as pre-adolescents with a maturity offset of −2.11 ± 0.68 years. Body Mass Index (BMI) was calculated by the ratio of body mass to the body standing height squared (19.6 ± 2.8 kg/m^2). All of them were healthy, with no musculoskeletal or neurological disease or lower limb injury. They joined two times a week for 90 min in a sports club, learning technical skills of team sports (soccer, handball, volleyball, basketball), in addition to the physical education class at school (according to school curriculum), two times per week for 45 min. Boys were asked to refrain from intense training 24 h prior to the testing days. Subjects' parents/legal guardians were informed about the experimental process and signed informed consent for the participation of their son/legal ward. The study was conducted according to the Declaration of Helsinki and was approved by the institutional research review board (EC-1/2020).

2.2. *Experimental Design/Procedures*

A randomized controlled design was used to investigate the acute effect of three warm-up protocols on 20-m sprint performance in preadolescent boys. The intervention protocols consisted of (a) a typical warm-up (control: CON), (b) 3 × 10-m maximal BwR bouts in addition to the typical warm-up, and (c) 3 × 10-m maximal FwR bouts added to the typical warm-up. Each of these protocols was assessed in random order, at three sessions carried out on non-consecutive days, separated by 72 h, at an indoor gym (wood parquet flooring), at a regular time of the day (14:00–16:00) in order to minimize any possible impact of testing time [34]. Each protocol lasted approximately 8–9 min. The participants wore light clothing and the same footwear during each session.

Rating of perceived exertion (RPE) was acquired immediately after the execution of each warm–up. The participants were tested in a 20-m sprint 4 min after the completion of each warm-up, in order to avoid fatigue [12,13,19]. Only one trial was performed since consecutive assessments could affect the performance of each subsequent sprint. The same investigator supervised all procedures and measurements.

One week before the first session, all participants were familiarized with the 20-m sprint and BwR [35]. Special attention was focused on the correct BwR technique, by means of demonstration and verbal feedback, following the guidelines of Uthoff et al. [32]. During the first session, anthropometric data of all participants were collected. A digital scale (BC-543, TANITA, Tokyo, Japan) and a stadiometer (Bodymeter 206, Seca, Ningbo, China) was used to measure body mass to the nearest 0.1 kg and body height (standing and seated) to the nearest 0.1 cm, respectively.

2.3. *Intervention*

The control condition of the typical warm–up (CON), lasted approximately 8 min and consisted of 3 min jogging at a low–medium tempo, followed by dynamic stretching exercises for the lower limbs (Table 1). More specifically, the first 7 exercises were performed for a 10-m distance and after the end of each exercise the participant walked back to the starting point. Dynamic stretching was preferred to static, to eliminate any potential adverse effect in performance [36].

Table 1. The dynamic stretching exercises performed for 10 m after the 3 min jogging.

1. Hip in.	3. Heel kicks	5. Side steps (1 per side)	7. Knee hugs
2. Hip out	4. Speed skips	6. Karaoke (1 per side)	8. Front leg swings (10 per leg)

The two other conditions consisted of three additional sets of 10 m maximal BwR or FwR sprints. Participants returned to the starting position running forward at a low pace. Subjects received verbal encouragement during BwR and FwR to ensure that the conditioning stimulus was maximal.

2.4. *20 m Sprint Test*

Sprint time over 5, 10 and 20 m was measured during the 20 m sprint (Table S1). For this purpose, three photocell timing gates (Witty Wireless Training Timer, Microgate, Bolzano, Italy) were placed at 5,

10, and 20 m. Photocells were adjusted to the pelvis height of each participant [37]. Participants were instructed to start after the verbal signal "ready, go". They stood on an upright stride stance position, with their preferred foot forward, placed on the starting line over a pressure pad. Timing started when participants' foot was detached from the pressure pad. The assessor ensured no false steps before starting and correct starting posture before the start. During the sprints verbal encouragement motivated for maximal effort. Sprint speed was analyzed for the distances 0–5, 5–10, 0–10, 10–20, and 0–20 m and was calculated by dividing the running distance by the time.

2.5. Rate of Perceived Exertion (RPE)

RPE was measured immediately after the completion of each intervention (Table S1), using the 10-degree Children's OMNI scale [38]. Participants replied to the question *"how tired do you feel?"*, while the investigator showed them the schematic OMNI scale. Participants had to declare the requisite exertion by indicating a number on the scale from 0 (not tired at all) to 10 (very, very tired). During the last two sessions at the sports club they were familiarized with the scale. This included a thorough description and explanation of the scale and responding to any questions or doubts that they had.

2.6. Statistical Analysis

All data are presented as means and standard deviations. The dependent variables were the sprint speed for the distances of 0–5, 5–10, 0–10, 10–20, and 0–20 m and the RPE. The Shapiro-Wilk test was used to confirm the normal distribution of the data (p-values ranging from 0.126 to 0.850 among all variables), and Levene's test for the equality of variances ($p = 0.368$–0.787). Furthermore, Mauchly's test was performed to confirm that the assumption for sphericity was satisfied ($p = 0.067$–0.641). One-way Analysis of Variance (ANOVA) for repeated measurements was used for the statistical assessment to examine the effect of warm-up protocol (three levels: CON, BwR and FwR). The level of significance α was set at 0.05. Statistically significant effects were assessed with the Scheffé's post-hoc test. The effect sizes were calculated using eta squared (η^2). The one sample t-test was used to examine the change in percent of sprint performance during the BwR or FwR relative to the CON condition compared to baseline zero. Confidence intervals at 95% confidence level ($CI_{95\%}$) were constructed. Statistical analysis was performed with SPSS for Windows, version 25 (IBM Corp., Armonk, NY, USA) and custom scripts in R, version 3.6.1 (R Foundation for Statistical Computing, Vienna, Austria).

3. Results

3.1. Sprint Speed

Sprint speed was not affected by protocol for the distances 0–5 m ($F(2,26) = 0.34$, $p > 0.05$, $\eta^2 = 0.03$), 5–10 m ($F(2,26) = 0.27$, $p > 0.05$, $\eta^2 = 0.04$), 0–10 m ($F(2,26) = 0.46$, $p > 0.05$, $\eta^2 = 0.10$) and 0–20 m ($F(2,26) = 0.79$, $p > 0.05$, $\eta^2 = 0.06$) (Table 2). However, a statistically significant effect of protocol on sprint speed was detected for the 10-20m distance ($F(2,26) = 5.85$, $p = 0.008$, $\eta^2 = 0.31$). More specifically, post-hoc tests revealed significantly higher sprint speed over the 10-20m distance after the BwR protocol compared to control ($p = 0.019$, $CI_{95\%}$: 0.025 to 0.30).

The percent change in sprint speed after the BwR and FwR relative to the CON protocol was highly variable among subjects for the distances 0–5 m, 5–10 m and 0–10 m, revealing participants with either lower or higher performance than the CON protocol (Figure 1). More systematic trends were observed for 10–20 m and 0–20 m distances. More specifically, for the distance 10–20 m the speed after the BwR protocol was 2.4 ± 2.9% higher than the CON and it was statistically different from zero ($CI_{95\%}$: 0.8 to 4.1% $p = 0.008$) while the increase by 1.6 ± 2.9% for the FwR compared to the CON was not significantly different from zero ($CI_{95\%}$: −0.1 to 3.3%, $p = 0.065$). Regarding the 0–20 m distance speed after the BwR was 0.9 ± 2.6% higher than the CON and 0.0 ± 2.4% after the FwR protocol. Both percentages were not significant from zero (BwR $CI_{95\%}$: −0.6 to 2.4 $p = 0.241$, and FwR $CI_{95\%}$: −1.4 to 1.4 $p = 0.972$, respectively).

Table 2. Mean and standard deviation (SD) values of sprint speed (m/s) for the 20 m sprint and its intermittent distances for the warm-up protocols (CON: typical warm–up; BwR: typical warm–up plus 3 × 10 m backward running bouts; FwR: typical warm–up plus 3 × 10 m forward running bouts). Significantly higher values compared to the CON protocol are designated with asterisks (*: $p < 0.01$).

Distance	CON	BwR	FwR	*p*-Value
0–5 m	4.64 ± 0.28	4.66 ± 0.38	4.58 ± 0.34	0.714
5–10 m	5.53 ± 0.45	5.47 ± 0.29	5.47 ± 0.37	0.769
0–10 m	5.04 ± 0.27	5.02 ± 0.28	4.98 ± 0.29	0.634
10–20 m	6.15 ± 0.50	6.31 ± 0.61 *	6.25 ± 0.56	0.008
0–20 m	5.53 ± 0.33	5.58 ± 0.37	5.54 ± 0.38	0.465

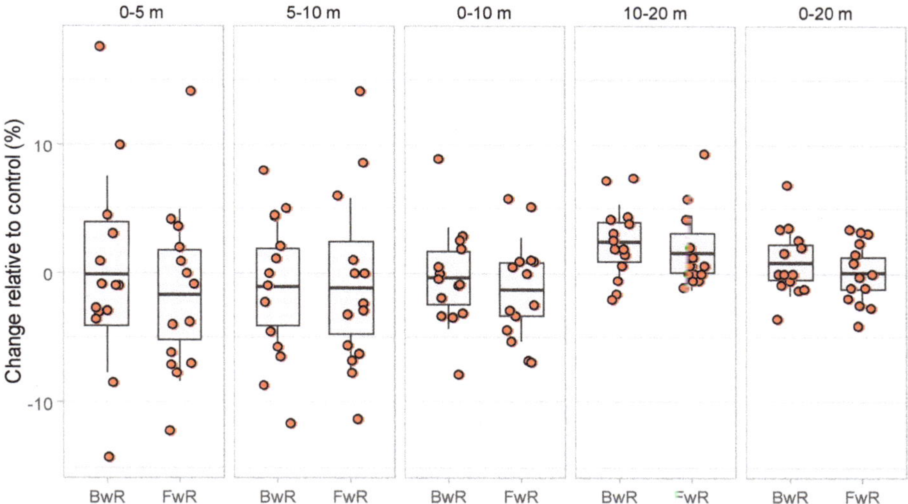

Figure 1. Percent change in 20 m sprint speed and its intermittent distances after the BwR and FwR protocols relative to the control protocol. Gray area corresponds to the $CI_{95\%}$ and vertical lines to the SD of mean, shown as a horizontal line in the middle of the gray area.

3.2. Rate of Perceived Exertion

A statistically significant effect of protocol was found on RPE ($F(2,26) = 24.2$, $p < 0.001$). Post-hoc tests revealed a statistically significantly lower RPE in the CON protocol (1.9 ± 0.8, $p < 0.001$) compared to the BwR (4.1 ± 1.5) and FwR (4.2 ± 1.2) warm-up protocols. This indicates that implementing either 3 × 10 m BwR or FwR after a typical warm-up causes a similar RPE, which is higher relative to the typical warm-up per se.

4. Discussion

Adding 3 × 10 m sprints after a regular warm-up, regardless of the direction of running (BwR or FwR), caused a higher RPE but no significant improvement in the 20 m sprint speed compared to the typical warm-up (CON). Similarly, no significant effect of the warm-up protocol was observed for all intermittent distances of the sprint, except for the 10–20 m, where only the BwR protocol was superior compared to the CON in terms of sprint speed. Although the initial hypothesis for improved performance in 20 m sprint speed after the BwR compared to the FwR or CON protocols was not confirmed, these findings may give some limited evidence, that BwR could be an alternative means for enhancing performance for certain phases of a sprint in preadolescent boys.

To the best of our knowledge, this is the first study to examine the acute effect of running stimuli on sprint performance in preadolescent boys. Thus, it is difficult to directly compare the results of this study to those of other PAPE studies because of methodological differences. However, the fact that in the present study BwR or FwR failed to trigger a PAPE effect on 20-m sprinting performance in preadolescent boys is in accordance with previous research, regarding squat jumps with maximal isometric half-squats as conditioning [19]. More specifically, Arabatzi et al. [19], showed that among adults, adolescents and preadolescents of both sexes, jumping performance improved only in adult males and not in the other age and sex groups. One possible reason for the absence of PAPE in children could be their muscle fiber distribution [39], which is possibly lower in fast-twitch muscle fiber content, that are more prone to post-activation potentiation [40]. Furthermore, the training level seems to play a crucial role for the appearance of PAPE [21], and athletes with a high level of power or strength show a greater PAPE effect than athletes with lower values of power or strength [41,42]. More particularly, regarding sprinting speed, increased muscle stiffness and improved capacity to use effectively the stretch-shortening cycle are two factors linked to sprint performance and might be affected by training [43,44]. On the other hand, children have more compliant musculotendinous system [45,46], and insufficiently use their stretch-shortening cycle [29,30]. Recent studies indicate training may improve the former [32] but not the latter [47] in young athletes (adolescents and preadolescents, respectively). Hence, the existence of an immature neuromuscular system might also explain why the children that participated in the current study, which were in principle untrained (recreationally active), had no significant improvement in their 20 m sprint speed.

Beyond the age and training level, there are some additional factors that might have influenced the amount of PAPE on the 20-m sprint speed after the tested protocols. One of these factors could be the conditioning stimulus properties. The great variability of the effect of the conditioning stimulus, especially during the first 10 m, suggests that the optimal conditioning stimulus should be individualized as proposed by previous researchers [6,48]. This possibly explains the absence of differences in sprint speed between the protocol during the first 10 m. Previous studies have suggested that the reason for no positive effect of explosive conditioning stimuli on PAPE in 11- to 13-year-old gymnasts could be the relatively low volume and intensity [20]. In the present study, the effort of the trials could not be further increased since it was maximal. However, although there are possibilities to increase the load on the muscles, by adding resistance during the sprint, studies in adults have shown that performing sprint with resistance (backward sled towing) as conditioning did not improve their sprint speed for the first 5 m [49], which is in agreement with the current study. Nonetheless, the increased RPE observed after the end of the BwR and FwR protocol should be considered when planning future studies because in the presence of fatigue, adverse effects in performance might be expected [50]. Therefore, attempts to further increase the number of repetitions or the total covered distance, or the resistance during running (e.g., elastic bands), might have adverse effects on performance. However, the optimal load to maximize performance is still unknown.

Moreover, sprints require anaerobic power [51], whereas children have a decreased capacity to utilize their anaerobic metabolism [52,53]. Furthermore, sprinting, as a multi-joint, complex, circular and dynamic motion [54], is a challenging task for untrained children that have limited capacity to coordinate and activate optimally their muscles during complex movements [30,55]. In agreement with other studies [56–58], the lack of lower limb neuromuscular coordination might also explain the greater variability in percent change among the tested protocols, especially during the first 10 m of the acceleration, when the coordination demands are higher [51]. It is possible though that a larger sample size (n > 14) could reduce the probability for type II error in the case of 0–10 m distance. Alternatively, the lower variability shown at the 10–20 m distance could reveal a statistical differentiation in the BwR compared to the CON warm-up protocol. Hence, not only the volume and intensity, but also the nature of the conditioning stimulus could also play a role on the absence or presence of a PAPE effect.

This was also the main purpose of the study, i.e., to evaluate the PAPE effect of two protocols with conditioning stimuli of different nature (BwR and FwR) compared to the CON condition. Indeed, for the distance 10–20 m, a warm-up including the BwR was superior in terms of sprint speed compared to the CON protocol (mean difference 0.16 m/s), whereas this was not the case for the FwR protocol. One explanation for this limited but statistically significant difference could be that BwR might be a better conditioning stimulus, since it involves more concentric contractions [28] and children are not able to execute eccentric movements involving the stretch-shortening cycle, as effectively as adults do [29,30]. However, considering that using eccentric contractions as conditioning stimulus is more effective than concentric [59], suggesting BwR as a means of inducing PAPE is still a compromise. Therefore, BwR could be suggested for novice athletes to improve their performance, but the main goal of the strength and conditioning trainer should be to improve their technique and performance using—among others—plyometric programs, which are effective in young ages [60,61].

Another factor that might contribute to the presence of PAPE, is the optimal timing between the end of the conditioning stimulus and the test [7,41]. Immediately after the end of the conditioning stimulus, fatigue may mask any PAPE effect [50]. The fact that in the present study the sprint speed after the BwR or FwR was not lower than the CON protocol, shows that despite the increased RPE values, a rest interval of 4 minutes after the conditioning was enough to maintain performance levels. Nonetheless, considering previous findings showing that children, compared to adults, recover faster, rely more on their aerobic mechanisms for energy production, and are more resistant to fatigue [53], it is reasonable to argue that shorter rest intervals might have the potential for greater PAPE in children. However, this requires further investigation.

Regarding RPE and metabolic cost, BwR at maximal intensity is considered to have greater energy consumption than FwR [26]. Nonetheless, in a recently published paper, RPE and metabolic cost during BwR and FwR, at self-pace speed, was similar [62]. Both of the previously mentioned studies involved adults. Considering the above, it could be assumed that one of the reasons why children had no significant difference in RPE between the BwR and FwR protocols in the present study, could be their potential inability to perform the task maximally. However, this assumption requires further investigation in the future to be verified.

From a practical point of view, the findings of this study support the inclusion of BwR sprints in warm-up routines in preadolescent children, as a method to improve sprint performance across 10–20 m distance. This acute effect in performance may enhance performance during training or competition. However, these findings regard recreationally active preadolescent children and cannot be generalized to the population of any specific sport. Each sport has different demands and the training stimuli may vary as well. Therefore, the existence and extent of improvement in sprinting velocity after BwR sprints, remains to be verified, for distances that are of specific interest to each sport.

5. Conclusions

Although the implementation of 3 × 10 m sprints, either BwR or FwR, to a warm-up does not enhance 20 m sprint speed in recreationally active preadolescent boys, after a recovery period of 4 min, the positive effect of BwR on sprint speed during the distance 10–20 m suggests that BwR might be an alternative means for enhancing performance in certain phases of a sprint speed. However, preadolescent boys' response to different sprint conditioning exercises, optimal rest time and/or conditioning stimuli remains to be determined on an individual basis, taking into account the basic characteristics and limitations of children's physiology.

Supplementary Materials: The following are available online at http://www.mdpi.com/2075-4663/8/4/55/s1, Table S1: Speed and RPE Values of All Participants.

Author Contributions: Conceptualization, D.P. and E.B.; Methodology, D.P., E.B., A.P. and D.A.P.; Validation, E.B., A.X and D.A.P.; Data curation, D.P., E.B. and A.P.; Formal analysis, E.B. and D.A.P.; Investigation, D.P., E.B. and A.P.; Visualization, D.A.P.; Writing—original draft, D.P., E.B., and D.A.P.; Writing—review & editing, D.P., E.B., A.P., A.X. and D.A.P.; Supervision, E.B.. All authors have read and agreed to the published version of the manuscript.

Funding: This research received no external funding.

Acknowledgments: The authors would like to thank the participants and their parents for their dedicated time and collaboration.

Conflicts of Interest: The authors have no conflict of interest to declare.

References

1. Bishop, D.J. Warm Up II. Performance changes following active warm up and how to structure the warm up. *Sport. Med.* **2003**, *33*, 483–498. [CrossRef]
2. Gray, S.R.; Soderlund, K.; Watson, M.; Ferguson, R.A. Skeletal muscle ATP turnover and single fibre ATP and PCr content during intense exercise at different muscle temperatures in humans. *Pflügers Arch. Eur. J. Physiol.* **2011**, *462*, 885–893. [CrossRef] [PubMed]
3. Racinais, S.; Oksa, J. Temperature and neuromuscular function. *Scand. J. Med. Sci. Sports* **2010**, *20*, 1–18. [CrossRef]
4. Jones, A.M.; Burnley, M. Oxygen uptake kinetics: An underappreciated determinant of exercise performance. *Int. J. Sports Physiol. Perform.* **2009**, *4*, 524–532. [CrossRef] [PubMed]
5. Wilson, J.M.; Duncan, N.M.; Marin, P.J.; Brown, L.E.; Loenneke, J.P.; Wilson, S.M.C.; Jo, E.; Lowery, R.P.; Ugrinowitsch, C. Meta-analysis of postactivation potentiation and power. *J. Strength Cond. Res.* **2013**, *27*, 854–859. [CrossRef] [PubMed]
6. McBride, J.M.; Nimphius, S.; Erickson, T.M. The acute effects of heavy-load squats and loaded countermovement jumps on sprint performance. *J. Strength Cond. Res.* **2005**, *19*, 893–897.
7. Kilduff, L.P.; Bevan, H.R.; Kingsley, M.I.C.; Owen, N.J.; Bennett, M.A.; Bunce, P.J.; Hore, A.M.; Maw, J.R.; Cunningham, D.J. Postactivation potentiation in professional rugby players: optimal recovery. *J. Strength Cond. Res.* **2007**, *21*, 1134. [CrossRef] [PubMed]
8. Cuenca-Fernández, F.; Smith, I.C.; Jordan, M.J.; Macintosh, B.R.; López-contreras, G.; Arellano, R.; Herzog, W. Nonlocalized postactivation performance enhancement (PAPE). *Appl. Physiol. Nutr. Metab.* **2017**, *1125*, 1122–1125. [CrossRef]
9. Sale, D.G. Postactivation potentiation: Role in human performance. *Exerc. Sport Sci. Rev.* **2002**, *30*, 138–143. [CrossRef]
10. Hodgson, M.J.; Docherty, D.; Robbins, D. Post-activation potentiation. *Sport. Med.* **2005**, *35*, 585–595. [CrossRef]
11. Blazevich, A.J.; Babault, N. Post-activation potentiation versus post-activation performance enhancement in humans: Historical perspective, underlying mechanisms, and current issues. *Front. Physiol.* **2019**, *10*, 1359. [CrossRef] [PubMed]
12. Chatzopoulos, D.E.; Michailidis, C.J.; Giannakos, A.K.; Alexiou, K.C.; Patikas, D.A.; Antonopoulos, C.B.; Kotzamanidis, C.M. Postactivation potentiation effects after heavy resistance exercise on running speed. *J. Strength Cond. Res.* **2007**, *21*, 1278–1281. [PubMed]
13. Rahimi, R. The acute effects of heavy versus light-load squats on sprint performance. *Phys. Educ. Sport* **2007**, *5*, 163–169.
14. Byrne, P.J.; Kenny, J.; O' Rourke, B. Acute pobloomtentiating effect of depth jumps on sprint performance. *J. Strength Cond. Res.* **2014**, *28*, 610–615. [CrossRef]
15. van den Tillaar, R.; Lerberg, E.; von Heimburg, E. Comparison of three types of warm-up upon sprint ability in experienced soccer players. *J. Sport Heal. Sci.* **2019**, *8*, 574–578. [CrossRef]
16. Yoshimoto, T.; Takai, Y.; Kanehisa, H. Acute effects of different conditioning activities on running performance of sprinters. *Springerplus* **2016**, *5*, 1203. [CrossRef]
17. Ferreira-Júnior, J.B.; Guttierres, A.P.M.; Encarnação, I.G.A.; Lima, J.R.P.; Borba, D.A.; Freitas, E.D.S.; Bemben, M.G.; Vieira, C.A.; Bottaro, M. Effects of different conditioning activities on 100-m dash performance in high school track and field athletes. *Percept. Mot. Skills* **2018**, *125*, 003151251876449. [CrossRef]

18. Pääsuke, M.; Ereline, J.; Gapeyeva, H. Twitch contraction properties of plantar flexor muscles in pre- and post-pubertal boys and men. *Eur. J. Appl. Physiol.* **2000**, *82*, 459–464. [CrossRef]
19. Arabatzi, F.; Patikas, D.A.; Zafeiridis, A.; Giavroudis, K.; Kannas, T.; Gourgoulis, V.; Kotzamanidis, C.M. The post-activation potentiation effect on squat jump performance: Age and sex effect. *Pediatr. Exerc. Sci.* **2014**, *26*, 187–194. [CrossRef]
20. Dallas, G. The post-activation effect with two different conditioning stimuli on drop jump performance in pre-adolescent female gymnasts. *Artic. J. Phys. Educ. Sport* **2018**, *18*, 2368–2374.
21. Tsimachidis, C.; Patikas, D.A.; Galazoulas, C.; Bassa, E.I.; Kotzamanidis, C.M. The post-activation potentiation effect on sprint performance after combined resistance/sprint training in junior basketball players. *J. Sports Sci.* **2013**, *31*, 1117–1124. [CrossRef] [PubMed]
22. Karcher, C.; Buchheit, M. On-court demands of elite handball, with special reference to playing positions. *Sport. Med.* **2014**, *44*, 797–814. [CrossRef] [PubMed]
23. Stojanović, E.; Stojiljković, N.; Scanlan, A.T.; Dalbo, V.J.; Berkelmans, D.M.; Milanović, Z. The activity demands and physiological responses encountered during basketball match-play: A systematic review. *Sport. Med.* **2018**, *48*, 111–135. [CrossRef] [PubMed]
24. Bloomfield, J.; Polman, R.; O'Donoghue, P. Physical demands of different positions in FA premier league soccer. *J. Sport. Sci. Med.* **2007**, *6*, 63–70.
25. Flynn, T.W.; Soutas-Little, R.W. Mechanical power and muscle action during forward and backward running. *J. Orthop. Sports Phys. Ther.* **1993**, *17*, 108–112. [CrossRef]
26. Wright, S.; Weyand, P.G. The application of ground force explains the energetic cost of running backward and forward. *J. Exp. Biol.* **2001**, *204*, 1805–1815.
27. Roos, P.E.; Barton, N.; van Deursen, R.W.M. Patellofemoral joint compression forces in backward and forward running. *J. Biomech.* **2012**, *45*, 1656–1660. [CrossRef]
28. Cavagna, G.A.; Legramandi, M.A.; La Torre, A. An analysis of the rebound of the body in backward human running. *J. Exp. Biol.* **2012**, *215*, 75–84. [CrossRef]
29. Gillen, Z.M.; Jahn, L.E.; Shoemaker, M.E.; Mckay, B.D.; Mendez, A.I.; Bohannon, N.A.; Cramer, J.T. Effects of eccentric preloading on concentric vertical jump performance in youth athletes. *J. Appl. Biomech.* **2019**, *35*, 327–335. [CrossRef]
30. Lazaridis, S.N.; Bassa, E.I.; Patikas, D.A.; Hatzikotoulas, K.; Lazaridis, F.K.; Kotzamanidis, C.M. Biomechanical comparison in different jumping tasks between untrained boys and men. *Pediatr. Exerc. Sci.* **2013**, *25*, 101–113. [CrossRef]
31. Meyers, R.W.; Oliver, J.L.; Hughes, M.G.; Cronin, J.B.; Lloyd, R.S. Maximal sprint speed in boys of increasing maturity. *Pediatr. Exerc. Sci.* **2015**, *27*, 85–94. [CrossRef] [PubMed]
32. Uthoff, A.; Oliver, J.; Cronin, J.; Harrison, C.; Winwood, P. Sprint-specific training in youth: Backward running vs. forward running training on speed and power measures in adolescent male athletes. *J. Strength Cond. Res.* **2020**, *34*, 1113–1122. [CrossRef] [PubMed]
33. Mirwald, R.L.; Baxter-Jones, A.D.G.; Bailey, D.A.; Beunen, G.P. An assessment of maturity from anthropometric measurements. *Med. Sci. Sports Exerc.* **2002**, *34*, 689–694. [PubMed]
34. Thun, E.; Bjorvatn, B.; Flo, E.; Harris, A.; Pallesen, S. Sleep, circadian rhythms, and athletic performance. *Sleep Med. Rev.* **2015**, *23*, 1–9. [CrossRef]
35. Uthoff, A.; Oliver, J.; Cronin, J.; Winwood, P.; Harrison, C. Prescribing target running intensities for high-school athletes: Can forward and backward running performance be autoregulated? *Sports* **2018**, *6*, 77. [CrossRef]
36. Behm, D.G.; Chaouachi, A. A review of the acute effects of static and dynamic stretching on performance. *Eur. J. Appl. Physiol.* **2011**, *111*, 2633–2651. [CrossRef]
37. Yeadon, M.R.; Kato, T.; Kerwin, D.G. Measuring running speed using photocells. *J. Sports Sci.* **1999**, *17*, 249–257. [CrossRef]
38. Robertson, R.J.; Goss, F.L.; Andreacci, J.L.; Dubé, J.J.; Rutkowski, J.J.; Snee, B.M.; Kowallis, R.A.; Crawford, K.; Aaron, D.J.; Metz, K.F. Validation of the children's OMNI RPE scale for stepping exercise. *Med. Sci. Sports Exerc.* **2005**, *37*, 290–298. [CrossRef]

39. Lexell, J.; Sjöström, M.; Nordlund, A.S.; Taylor, C.C. Growth and development of human muscle: A quantitative morphological study of whole vastus lateralis from childhood to adult age. *Muscle Nerve* **1992**, *15*, 404–409. [CrossRef]
40. Hamada, T.; Sale, D.G.; MacDougall, J.D.; Tarnopolsky, M.A. Interaction of fibre type, potentiation and fatigue in human knee extensor muscles. *Acta Physiol. Scand.* **2003**, *178*, 165–173. [CrossRef]
41. Jo, E.; Judelson, D.A.; Brown, L.E.; Coburn, J.W.; Dabbs, N.C. Influence of recovery duration after a potentiating stimulus on muscular power in recreationally trained individuals. *J. Strength Cond. Res.* **2010**, *24*, 343–347. [CrossRef]
42. Smilios, I.; Sotiropoulos, K.; Barzouka, K.; Christou, M.; Tokmakidis, S.P. Contrast loading increases upper body power output in junior volleyball athletes. *Pediatr. Exerc. Sci.* **2017**, *29*, 103–108. [CrossRef] [PubMed]
43. Ross, A.; Leveritt, M.; Riek, S. Neural influences on sprint running: training adaptations and acute responses. *Sport. Med.* **2001**, *31*, 409–425. [CrossRef] [PubMed]
44. Meyers, R.W.; Oliver, J.L.; Hughes, M.G.; Lloyd, R.S.; Cronin, J.B. The influence of maturation on sprint performance in boys over a 21-month period. *Med. Sci. Sports Exerc.* **2016**, *48*, 2555–2562. [CrossRef] [PubMed]
45. Kubo, K.; Teshima, T.; Hirose, N.; Tsunoda, N. Growth changes in morphological and Mechanical properties of human patellar tendon In Vivo. *J. Appl. Biomech.* **2014**, *30*, 415–422. [CrossRef] [PubMed]
46. Waugh, C.M.; Blazevich, A.J.; Fath, F.; Korff, T. Age-related changes in mechanical properties of the Achilles tendon. *J. Anat.* **2012**, *220*, 144–155. [CrossRef]
47. Katsikari, K.; Bassa, E.I.; Skoufas, D.; Lazaridis, S.N.; Kotzamanidis, C.; Patikas, D.A. Kinetic and kinematic changes in vertical jump in prepubescent girls after 10 weeks of plyometric training. *Pediatr. Exerc. Sci.* **2020**, *32*, 1–8. [CrossRef]
48. Till, K.A.; Cooke, C. The effects of postactivation potentiation on sprint and jump performance of male academy soccer players. *J. Strength Cond. Res.* **2009**, *23*, 1960–1967. [CrossRef]
49. Monaghan, D.J.; Cochrane, D.J. Can backward sled towing potentiate sprint performance? *J. Strength Cond. Res.* **2019**, *34*, 1. [CrossRef]
50. Tillin, N.A.; Bishop, D.J. Factors modulating post-activation potentiation and its effect on performance of subsequent explosive activities. *Sport. Med.* **2009**, *39*, 147–166. [CrossRef]
51. Piechota, K.; Pakosz, P.; Borysiuk, Z.; Wa, Z. Coordination aspects of an effective sprint start. *Front. Physiol.* **2018**, *9*, 1–7.
52. Kaczor, J.J.; Ziolkowski, W.; Popinigis, J.; Tarnopolsky, M.A. Anaerobic and aerobic enzyme activities in human skeletal muscle from children and adults. *Pediatr. Res.* **2005**, *57*, 331–335. [CrossRef] [PubMed]
53. Ratel, S.; Blazevich, A.J. Are prepubertal children metabolically comparable to well-trained adult endurance athletes? *Sport. Med.* **2017**, *47*, 1477–1485. [CrossRef] [PubMed]
54. Cochrane, D. The sports performance application of vibration exercise for warm-up, flexibility and sprint speed. *Eur. J. Sport Sci.* **2013**, *13*, 256–271. [CrossRef]
55. Lazaridis, S.N.; Bassa, E.I.; Patikas, D.A.; Giakas, G.; Gollhofer, A.; Kotzamanidis, C.M. Neuromuscular differences between prepubescents boys and adult men during drop jump. *Eur. J. Appl. Physiol.* **2010**, *110*, 67–74. [CrossRef]
56. Cortis, C.; Tessitore, A.; Perroni, F.; Lupo, C.; Pesce, C.; Ammendolia, A.; Capranica, L. Interlimb coordination, strength, and power in soccer players across the lifespan. *J. Strength Cond. Res.* **2009**, *23*, 2458–2466. [CrossRef]
57. Cortis, C.; Tessitore, A.; Lupo, C.; Pesce, C.; Fossile, E.; Figura, F.; Capranica, L. Inter-limb coordination, strength, jump, and sprint performances ollowing a youth men's basketball game. *J. Strength Cond. Res.* **2011**, *25*, 135–142. [CrossRef]
58. Lupo, C.; Ungureanu, A.N.; Varalda, M.; Brustio, P.R. Running technique is more effective than soccer-specific training for improving the sprint and agility performances with ball possession of prepubescent soccer players. *Biol. Sport* **2019**, *36*, 249–255. [CrossRef]
59. Seitz, L.B.; Haff, G.G. Factors modulating post-activation potentiation of jump, sprint, throw, and upper-body ballistic performances: A systematic review with meta-analysis. *Sport. Med.* **2016**, *46*, 231–240. [CrossRef]
60. Johnson, B.A.; Salzberg, C.L.; Stevenson, D.A. A systematic review: Plyometric training programs for young children. *J. Strength Cond. Res.* **2011**, *25*, 2623–2633. [CrossRef]

61. Chaouachi, A.; Ben, O.A.; Hammami, R.; Drinkwater, E.J.; Behm, D.G. The combination of plyometric and balance training improves sprint and shuttle run performances more often than plyometric-only training with children. *J. Strength Cond. Res.* **2014**, *28*, 401–412. [CrossRef] [PubMed]
62. Masumoto, K.; Galor, K.; Craig-Jones, A.; Mercer, J.A. Metabolic costs during backward running with body weight support. *Int. J. Sports Med.* **2019**, *40*, 269–275. [CrossRef] [PubMed]

Article

Verifying Physiological and Biomechanical Parameters during Continuous Swimming at Speed Corresponding to Lactate Threshold

Gavriil G. Arsoniadis [1], Ioannis S. Nikitakis [1], Petros G. Botonis [1], Ioannis Malliaros [1] and Argyris G. Toubekis [1,2,*]

1 Division of Aquatic Sports, School of Physical Education and Sports Science, National and Kapodistrian University of Athens, Dafne, 17237 Athens, Greece; garsoniadis@phed.uoa.gr (G.G.A.); inikitak@phed.uoa.gr (I.S.N.); pboton@phed.uoa.gr (P.G.B.); gmalliaros@phed.uoa.gr (I.M.)

2 Sports Performance Laboratory, School of Physical Education and Sport Science, National and Kapodistrian University of Athens, Dafne, 17237 Athens, Greece

* Correspondence: atoubekis@phed.uoa.gr; Tel.: +030-210-727-276-049

Received: 3 May 2020; Accepted: 30 June 2020; Published: 30 June 2020

Abstract: The purpose of this study was to verify the physiological responses and biomechanical parameters measured during 30 min of continuous swimming (T30) at intensity corresponding to lactate threshold previously calculated by an intermittent progressively increasing speed test (7 × 200 m). Fourteen competitive swimmers (18.0 (2.5) years, 67.5 (8.8) kg, 174.5 (7.7) cm) performed a 7 × 200 m front crawl test. Blood lactate concentration (BL) and oxygen uptake (VO_2) were determined after each 200 m repetition, while heart rate (HR), arm-stroke rate (SR), and arm-stroke length (SL) were measured during each 200 m repetition. Using the speed vs. lactate concentration curve, the speed at lactate threshold (sLT) and parameters corresponding to sLT were calculated (BL-sLT, VO_2-sLT, HR-sLT, SR-sLT, and SL-sLT). In the following day, a T30 corresponding to sLT was performed and BL-T30, VO_2-T30, HR-T30, SR-T30, and SL-T30 were measured after the 10th and 30th minute, and average values were used for comparison. VO_2-sLT was no different compared to VO_2-T30 ($p > 0.05$). BL-T30, HR-T30, and SR-T30 were higher, while SL-T30 was lower compared to BL-sLT, HR-sLT, SR-sLT, and SL-sLT ($p < 0.05$). Continuous swimming at speed corresponding to lactate threshold may not show the same physiological and biomechanical responses as those calculated by a progressively increasing speed test of 7 × 200 m.

Keywords: lactate threshold; continuous swimming; physiological responses; biomechanical parameters; validity

1. Introduction

Progressive discontinuous swim protocols, such as a 7 × 200 m progressively increasing speed test, are commonly used to evaluate both physiological [1,2] and biomechanical [3] characteristics in swimming. More specifically, a 7 × 200 m test is used to identify aerobic training intensity domains and subsequent changes during a year-round training plan [4]. The identification of training intensity domains requires drawing a speed vs. blood lactate concentration curve and calculating specific aerobic indices, such as speed corresponding to first and second lactate thresholds [5,6]. The sLT (speed at lactate threshold) is one of the most frequently used indices to assess swimming endurance capacity [1,7], and several methods are utilized for its calculation [8].

In swimming, the most frequently used method for sLT calculation is x-axis projection of the intersection of two lines connecting the three higher and four lower points of the speed lactate curve [7]. Subsequently, biomechanical or physiological parameters corresponding to sLT may be calculated to provide additional information for coaches (i.e., arm-stroke rate (SR), arm-stroke length (SL), heart rate (HR), and blood lactate concentration (BL) corresponding to sLT). However, all suggested methods used for sLT calculation present errors in estimation, and this may be transferred to the training pace prescription of swimmers [9]. The validity of sLT is tested by calculating the speed corresponding to the maximum lactate steady state (MLSS: maximum lactate concentration that can be maintained constant during continuous exercise; [10,11]), which is a time-consuming test for verification and implies that sLT may be used during continuous swimming training. Whatever the case, there is a need to verify the calculated sLT and corresponding physiological and biomechanical variables obtained after a 7 × 200 m test during a continuous swimming training set so as to increase precision in the control of the training load and improve swimming performance [7,12]. A 30 min duration is an appropriate and acceptable time limit to compare variations in biomechanical and physiological parameters in continuous swimming [7]. Using a prescribed sLT speed, a specific response in physiological (HR, BL) or biomechanical variables (SR, SL) is expected, as these parameters are interconnected [13]. Therefore, verifying this information is important before coaches plan a training set. However, to our knowledge, there has been no previous study to verify the calculated sLT using a continuous 30 min of swimming.

Thus, the purpose of the current study was to verify the physiological responses and biomechanical parameters during continuous swimming at intensity corresponding to the lactate threshold previously calculated by an intermittent progressively increasing speed test (7 × 200 m). We hypothesized that the calculated parameters would be verified during the continuous swimming effort.

2. Materials and Methods

2.1. Participants

Fourteen regional and national level male and female swimmers (ten sprinters and four middle-distance swimmers) that specialized in competitive distances of 100, 200, and 400 m volunteered to participate in the study (Table 1). Participants had a training background of 9.9 (1.8) years and they participated in daily training (6 days per week) with a duration of approximately two hours per session. Participants were randomly selected from local swimming clubs after getting agreement from parents and coaches. Swimmers were not consuming nutritional supplements during the testing period; they were asked to consume the same diet two days before the trials. Participants were instructed to avoid alcohol or caffeine consumption two days before each testing session. Each participant and his or her legal guardian provided written informed consent after receiving thorough explanation of the study. The local institutional review board approved the experimental procedures (approval no: 1029/6/12/2017) in accordance with Helsinki declaration for human subjects.

Table 1. Anthropometrics and performance characteristics of competitive swimmers. The data are presented as mean values and standard deviation (*SD*) for both males and females and for each gender separately.

Variables	*N* = 14 Swimmers (Male and Female)	*N* = 10 Male Swimmers	*N* = 4 Female Swimmers
Age (years)	18.0 (2.5)	17.5 (2.4)	19.2 (2.6)
Body mass (kg)	67.5 (8.8)	69.7 (8.1)	60.3 (8.7)
Height (cm)	174.5 (7.7)	175.6 (7.3)	171.9 (9.3)
Time (s) 200 m front crawl	131.6 (7.3)	129.2 (7.3)	137.6 (2.5)
FINA points 200 m front crawl	513.6 (66.0)	503.3 (75.0)	539.5 (28.2)
FINA points (best style)	564.0 (109.3)	540.4 (119.4)	623.0 (49.5)
Competitive experience (years)	9.9 (1.8)	9.5 (1.9)	10.8 (1.5)

FINA: Fédération Internationale de NatationAmateur.

2.2. Study Design

Physiological and biomechanical parameters calculated during a progressively increasing speed swimming test (mean overall time: ~45 min) were compared to those measured during 30 min of continuous swimming. Swimmers were tested in two sessions 48 h apart (Figure 1), and all tests were completed at the same time of the day for each swimmer (between 14:00 to 16:00). Prior to the first testing session, body mass and height were measured (Seca, Hamburg, Germany). One day prior to the first testing session, as well as during the two days separating the two testing sessions, swimmers participated in an easy (BL: ~2 mmol L^{-1}) low volume endurance training (~3000 to 4000 m). The study was conducted during the specific preparation mesocycle of training. All swimming tests were performed using front crawl in a 25 m indoor swimming pool with a constant water temperature of 25 °C to 26 °C and 60% ambient humidity.

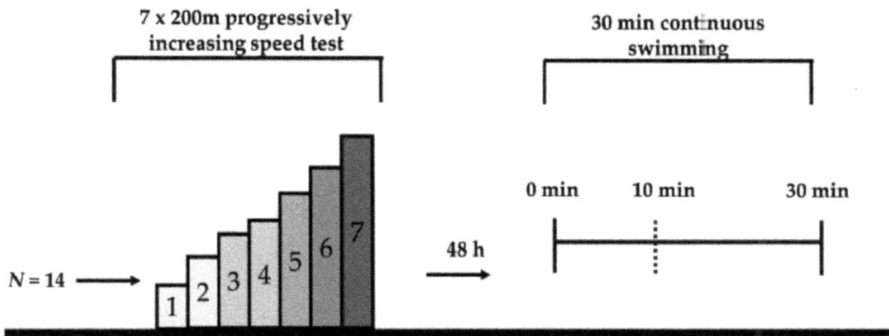

Figure 1. Experimental design of the current study; 7 × 200 m: 7 repetitions of 200 m front crawl, h: hours.

2.3. Progressively Increasing Swimming Speed Test and Parameters Calculation

All swimmers participated in a standardized swimming warm-up, and 10 min later performed a 7 × 200 m front crawl test at intensities calculated using the most recent 200 m race time and corresponding to 60%, 70%, 75%, 80%, 85%, 90%, and maximum effort. Participants were familiarized with the pace of the first two repetitions during a previous training session. During testing, one of the experimenters walked alongside the swimming pool providing guidance during each 200 m repetition. During the 7 × 200 m test, each repetition started every five minutes and 30 s with a push-off start from within the water [1]. Fingertip blood samples were collected after each repetition and were analyzed for BL (Lactate Scout$^+$, SensLab GmbH, Leipzig, Germany). Immediately after the completion of each 200 m repetition, a mouthpiece and a nose clip were attached to swimmers during recovery. Expired air collected during the first 20 s of recovery was analyzed for oxygen uptake (VO_2) using a portable gas analyzer (VO2000, Med Graphics, Saint Paul, MN, USA.; [14]). HR was recorded continuously using telemetry (s610i; Polar Electro, Kempele, Finland). A vest was used to keep the transmitter attached to each swimmer's chest while swimming. sLT and the respective BL-sLT were calculated by the x-axis projection of the intersection of lines connecting the three higher and four lower points of the speed lactate curve (BL-sLT: mean R^2 = 0.97 (0.02), mean r = 0.99 (0.01); 1). VO_2-sLT and HR-sLT were calculated by interpolation using linear regression of swimming speed versus VO_2 (mean R^2 = 0.99 (0.02), mean r = 0.99 (0.01)) or HR (mean R^2 = 0.99 (0.01), mean r = 0.99 (0.01). During the 7 × 200 m test, SR was calculated by the time (T) used to complete three arm-stroke cycles (180·T^{-1}) and SL was calculated by dividing swimming speed every 50 m (V) by SR. The time for three arm-stroke cycles was recorded using a handheld chronometer. SR-sLT and SL-sLT were calculated by interpolation using the best fit regression line of SR and SL versus swimming speed during the 7 × 200 m test (SR: mean R^2 = 0.98 (0.02), mean r = 0.99 (0 01); SL: mean R^2 = 0.96 (0.04), mean r = 0.98 (0.02)).

2.4. Continuous Swimming Session with Constant Speed

The swimmers participated in a continuous swimming session 48 h after the completion of the 7 × 200 m test. Continuous swimming speed session was completed after a standardized warm-up that consisted of 1000 m of swimming (400 m slow swimming at 60% intensity, 4 × 50 m front crawl kicks, 4 × 50 m front crawl drills, and 4 × 50 m front crawl swim with progressively increasing speed). Ten minutes after warm-up, the swimmers started a T30 at a constant speed corresponding to sLT. During T30, the swimmers kept the individual sLT speed constant while guided by a sound signal emitted by a transmitter placed next to the ear and under the swimming cap (FINIS tempo pro, Finis Inc., Livermore, CA, USA). The swimmers were instructed to adjust their speed in order to touch the wall at each sound signal. Additionally, one of the experimenters recorded the time for each 50 m split (HS-80; CASIO, Guangzhou, China). BL-T30 and VO_2-T30 were measured after the 10th and 30th minute of T30, while HR-T30 was recorded continuously. SR-T30 and SL-T30 were measured every 50 m during the T30 continuous swimming session. The 10th min was used to identify variations in physiological adjustments compared to the 30th min. Nevertheless, the mean values of BL, HR, SR, SL, and speed (s-T30) measured during the T30 were used for the statistical analysis.

2.5. Statistical Analysis

The student *t*-test for dependent samples was used to examine the differences in physiological and biomechanical parameters calculated during the 7 × 200 m progressively increasing speed test and those measured in T30. Specifically, comparisons between sLT, BL-sLT, VO_2-sLT, HR-sLT, SR-sLT, SL-sLT vs. s-T30, BL-T30, VO_2-T30, HR-T30, SR-T30, SL-T30, respectively, were applied. Pearson *r* correlation coefficient was used to examine the relationship between the calculated values from a 7 × 200 m test with that measured during T30. The effect size for paired comparisons using the pooled standard deviation as denominator was calculated with Cohen's *d* [15]. The effect size was considered trivial if the absolute value of Cohen's *d* was less than 0.20, small if it was between 0.20 and 0.50, medium if it was between 0.50 and 0.80, and large if it was greater than 0.80. The 95% confidence limits (95% CL) were also calculated for the mean differences between parameters. G-Power 3.1.9.4 software [16] was used to examine the power of analysis. Considering the sample size in the current study ($N = 14$), an *ES* of 0.80 was required to get a statistical power value greater than 0.80. For the estimation of agreement between parameters, Bland and Altman plots were used [17]. SPSS software (v.23, SPSS Inc., Chicago, IL, USA) was used for data analysis. Data are presented as mean and standard deviation (SD). Statistical significance was set at $p < 0.05$.

3. Results

3.1. Swimming Speed between Tests

sLT calculated after the 7 × 200 m test was similar with s-T30 (sLT: 1.317 (0.078) vs. s-T30: 1.316 (0.082) m·s^{-1}, mean difference (*SD*): −0.002 (0.010) m·s^{-1}, 95% CL: −0.008, 0.005·m·s^{-1}, $d = -0.02$, $p = 0.634$). A significant correlation was observed between sLT and s-T30 ($R^2 = 0.97$, $r = 0.98$, $p = 0.001$).

3.2. Comparison of Physiological Variables between Tests

Measured BL-T30 was higher compared to BL-sLT (BL-T30: 4.7 (2.3) vs. BL-sLT: 3.4 (0.8) mmol·L^{-1}, mean difference (*SD*): 1.3 (2.4) mmol·L^{-1}, 95% CL: 0.00, 2.35 mmol·L^{-1}, $d = 0.83$, $p = 0.05$; Figure 2), and these variables were not correlated ($R^2 = 0.001$, $r = 0.03$, $p = 0.92$). A Bland and Altman plot indicated agreement between calculated and measured values (Figure 2). During continuous swimming, HR-T30 was higher compared to HR-sLT (HR-T30: 173 (8) vs. HR-sLT: 161 (10) b·min^{-1}, mean difference (*SD*): 11 (11) b·min^{-1}, 95% CL: 6, 18 b·min^{-1}, $d = 1.24$, $p = 0.02$; Figure 2), and these variables were not correlated ($R^2 = 0.06$, $r = 0.25$, $p = 0.39$). Agreement between calculated HR-sLT and measured HR-T30 values was observed (Figure 2). Moreover, VO_2-T30 was not different compared to VO_2-sLT (VO_2-T30: 41.7 (6.8) vs. VO_2-sLT: 42.8 (6.2) ml·kg^{-1}·min^{-1}, mean difference (*SD*): −0.88 (4.5) ml·kg^{-1}·min^{-1}, 95%

CL: −3.22, 1.45 ml·kg^{-1}·min^{-1}, d = −0.16, p = 0.43) and these parameters were correlated (R^2 = 0.53, r = 0.73, p = 0.00). A Bland and Altman plot showed agreement between calculated VO_2-sLT and measured VO_2-T30 values (Figure 2).

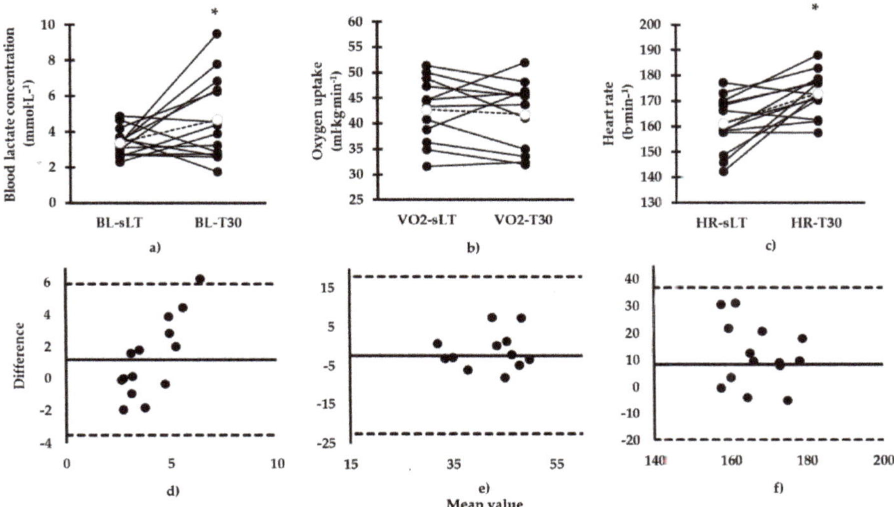

Figure 2. Comparison between calculated and measured parameters in the two tests, (**a**) blood lactate concentration (BL), (**b**) oxygen uptake (VO_2) (N = 12), (**c**) heart rate (HR). Individual values corresponding to BL-sLT, VO_2-sLT, and HR-sLT are compared with BL-T30, VO_2-T30, and HR-T30, respectively. Bland and Altman plots of mean vs. difference of two tests is presented in (**d**) BL-sLT and BL-T30, (**e**) VO_2-sLT and VO_2-T30, (N = 12), and (**f**) HR-sLT and HR-T30. Units of measure in (**d**), (**e**), and (**f**) are not shown for clarity and are the same as in the corresponding figure (**a**), (**b**), and (**c**) panels. * $p < 0.05$ between BL-sLT and BL-T30, VO_2-sLT and VO_2-T30, and between HR-sLT and HR-T30. sLT: corresponding to speed at lactate threshold; T30: 30 min of continuous swimming.

3.3. Comparison of Biomechanical Variables between Tests

During T30, the measured SR-T30 was higher compared to SR-sLT (SR-T30: 33.8 (2.9) vs. SR-sLT: 29.7 (4.1) cycles·min^{-1}, mean difference (SD): 4.0 (3.9) cycles·min^{-1}, 95% CL: 1.97, 6.07 cycles·min^{-1}, d = 1.15, p = 0.002; Figure 3). SR-T30 and SR-sLT were not correlated (R^2 = 0.16, r = 0.41, p = 0.15). A Bland and Altman plot indicated agreement between calculated SR-sLT and measured SR-T30 values (Figure 3). Measured SL-T30 was lower compared to SL-sLT (SL-T30: 2.3 (0.3) vs. SL-sLT: 2.6 (0.4) m·cycles^{-1}, mean difference (SD): −0.3 (0.2) m·cycles^{-1}, 95% CL: −0.40, −0.14 m·cycles^{-1}, d = −0.88, p = 0.001; Figure 3) and these parameters were correlated (R^2 = 0.55, r = 0.74, p = 0.003). A Bland and Altman plot showed agreement between calculated SL-sLT and measured SL-T30 values (Figure 3).

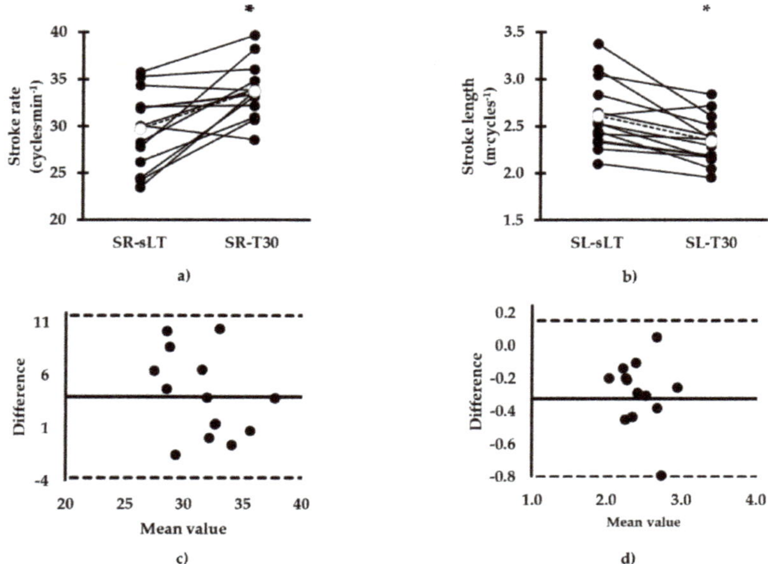

Figure 3. Comparison between calculated and measured parameters in the two tests, (**a**) comparison of arm-stroke rate (SR), (**b**) comparison of arm-stroke length (SL). Individual values corresponding to SR-sLT and SL-sLT are compared with SR-T30 and SL-T30, respectively. Bland and Altman plots of mean vs. difference of the two tests is presented in (**c**) SR-sLT vs. SR-T30, (**d**) SL-sLT vs. SL-T30. Units of measure in (**c**) and (**d**) are not shown for clarity and are the same as in the corresponding (**a**) and (**b**) panel. * $p < 0.05$ between SR-sLT and SR-T30 and between SL-sLT and SL-T30. sLT: corresponding to speed at lactate threshold; T30: 30 min of continuous swimming.

4. Discussion

The purpose of the current study was to verify the physiological and biomechanical parameters measured during continuous constant speed swimming corresponding to lactate threshold with those calculated after a 7 × 200 m test. The calculated sLT was successfully maintained during a T30 session. Calculated VO_2 was similar to measured VO_2, while a higher BL, HR, and SR and lower SL were recorded during T30. Bland and Altman plots indicated agreement, although a great bias was observed for all the physiological and biomechanical parameters.

A similar speed compared to sLT was expected in T30 on account of our experimental design. This is because swimmers were guided to follow a constant speed using audio signals. While maintaining the required speed during T30, some physiological adjustments were made by the swimmers. Increased BL and HR were observed during T30, and several factors may have contributed to this increment. First, we should examine the validity of sLT calculation with the method used in the present study. Previous studies report that using the intersection of two lines provides a speed corresponding to lactate threshold similar to MLSS [7]. However, the number of repetitions, the speed increment, and the duration of each repetition may influence the calculation of sLT [7]. Indeed, several methods, mathematical models, and various discontinuous protocols may indicate a different lactate threshold, which is not always similar to MLSS [9,18,19]. In fact, a methodological error of 2.0–2.5% in MLSS calculation should be considered [20]. This is because speed increments of this range are normally used when sLT is compared to MLSS [21]. In this case, a 2.0–2.5% lower speed compared to sLT may induce lower BL in T30, similar to sLT lactate values in the present study. Considering the above, a 2.5% error in calculating sLT should be expected even in studies reporting a valid estimation.

A second factor that needs to be mentioned is that lactate threshold as well as MLSS may present lactate concentration at a range of 2 to 8 mmol·L^{-1} [22]. In such a case, the swimmers in the current study showed increased lactate values in sLT but it is likely that they were still below or at their MLSS. Despite the limitation that MLSS was not measured, to allow a better understanding of BL-sLT and BL-T30 differences in the present study, it is expected that metabolic/physiological characteristics determine an athlete's ability to sustain a long duration effort. Endurance athletes are more efficient at maintaining long duration efforts with lower BL as opposed to sprint-oriented athletes at comparable relative exercise intensity [23]. Supporting this, Skorski et al. [24] found a 6.3% to 7.3% higher BL response in short-distance competitive swimmers during training sets such as 5 × 400 m or 5 × 200 m with constant speed corresponding to lactate threshold, suggesting that sLT may induce variable BL response during endurance training sets. In the present study, participants were mainly sprint-oriented and showed 27.3% higher BL than expected during continuous swimming in T30. In this case, the calculated sLT was probably not representing their steady physiological conditions leading to increased lactate production. Moreover, swimmers may show difficulty in maintaining a constant speed for more than 20 min, especially if they are not accustomed to do so, but may be able to maintain the same speed for longer durations in interval training set [25]. Nonetheless, a previous study also found a slightly higher BL during continuous swimming than predicted by 7 × 400 m or 7 × 200 m tests [7]. It seems that continuous effort may correspond to higher exercise intensity, as it has confirmed by higher physiological responses or by the inability to sustain constant speed for a long period [25].

A higher HR was observed during T30, indicating a higher effort during continuous swimming. A higher HR was measured in continuous exercise compared to that calculated corresponding to lactate threshold after a 7 × 200 m test in a previous study [7]. In the above study, increased swimming distance was accompanied by higher HR despite maintaining similar speed [7]. However, in the current study, some of the swimmers may not have reached a steady HR in all stages of the 7 × 200 m test because of the short time needed to complete 200 m repetitions, especially in the last stages (i.e., 130–160 s), thus underestimating the predicted sLT-HR value. Confirming the above information, Fernandes et al. [7] reported 2% (4 b·min^{-1}) higher HR corresponding to lactate threshold using 400 m compared to 200 m stages [7]. However, in the current study, a greater HR difference (11 b·min^{-1}) was observed between HR-sLT and HR-T30, possibly attributed to the training status and specialty of the swimmers. Additionally, a likely HR drift towards the last minutes of exercise attributed to cardiovascular adjustments during the long exercise duration in T30 cannot be excluded [26]. In contrast to BL and HR, VO_2-T30 was no different compared to VO_2-sLT. It has been indicated that oxygen uptake reaches values between ~80–100% of VO_{2peak} at the end of an endurance training set with a duration of 15 to 30 min [25,27]. Specifically, Pelarigo et al. [27] found constant VO_2 values that were ~85% of VO_{2max}, similar to the current study (85.5%). A combination of steady VO_2 response and increased BL during a continuous exercise may be observed during continuous efforts or in very heavy exercise intensity domains [28].

SR-T30 was increased whereas SL-T30 was decreased compared to SR-sLT and SL-sLT, respectively. Such changes are associated with an increased energy cost [29]. It is possible that the swimmers managed to adjust the applied force during each arm-stroke by increasing the relative duration of propulsive phases in order to maintain the required speed [30]. Similar results have been reported in a previous study in which less experienced athletes presented a decrease in SL with a concomitant increase in SR, despite swimming at a higher speed (by 2.5%) compared to MLSS [31]. On the contrary, Dekerle et al. [32] reported stability in SL during metabolically steady conditions in well-trained competitive swimmers. However, SL decreased at speeds above lactate threshold [21]. It seems that swimmers in the present study were exercising slightly above steady metabolic conditions, and then they were forced to alter their mechanics to maintain the required speed. These alterations in mechanics aim to overcome hydrodynamic drag and may lead to increments in metabolic response.

The abovementioned differences in physiological and biomechanical parameters indicate that the calculated parameters may not always correspond to the measured values during continuous swimming. However, this is in contrast to the observed agreement between BL-sLT, VO_2-sLT, HR-sLT, SR-sLT, and SL-sLT and BL-T30, VO_2-T30, HR-T30, SR-T30, and SL-T30, respectively, as indicated by Bland and Altman plots and has also been confirmed previously in a homogenous group of female middle- and long-distance swimmers [33]. Despite the observed agreements presented in the current study, the range of physiological and biomechanical differences observed is great. In this case, we cannot accept that calculated physiological and biomechanical parameters obtained by the intermittent protocol used in this study can predict corresponding ones during continuous long duration swimming. We should consider that BL-sLT, VO_2-sLT, and HR-sLT as well as SR-sLT and SL-sLT were calculated by equations obtained by the best fit of these parameters versus swimming speed, thus reducing the error of calculation. However, this was not reflected in the measured values, indicating that the observed differences represent real physiological and biomechanical gaps between predicted and measured variables. Further research may examine various mathematical models for lactate threshold calculation in swimmers.

5. Conclusions

The physiological and biomechanical parameters calculated by a progressively increasing swimming speed test corresponding to sLT may not be verified during continuous 30 min swimming in sprint and middle-distance swimmers. Swimmers maintaining constant speed corresponding to the second lactate threshold in a long duration 30 min swimming present individualized biomechanical and physiological adjustments that may not reflect the expected responses. In this case, an additional test for verification or a different mathematical model of lactate threshold calculation may be required to provide a valid guidance of training pace. Coaches should be aware that the individual data obtained by a progressively increasing speed test should be examined thoroughly and tested in training practice before planning a training set.

Author Contributions: Conceptualization: A.G.T. and G.G.A.; methodology: A.G.T., P.G.B., and G.G.A.; software: G.G.A.; validation: A.G.T. and P.G.B.; formal analysis: P.G.B.; investigation: G.G.A., I.S.N., and I.M.; resources: A.G.T.; data curation: G.G.A.; writing—original draft preparation: G.G.A.; writing—review and editing: A.G.T.; visualization: P.G.B.; supervision: A.G.T.; project administration: A.G.T. All authors have read and agreed to the published version of the manuscript.

Funding: This research received no external funding.

Acknowledgments: The authors would like to thank the swimmers for their participation in the study.

Conflicts of Interest: The authors declare no conflict of interest, financial or otherwise.

List of Abbreviations

7 × 200 m:	Progressively increasing speed test
T30:	30 min of continuous swimming
sLT:	Speed corresponding to the second lactate threshold
BL:	Blood lactate concentration
VO_2:	Oxygen uptake
HR:	Heart rate
SR:	Arm-stroke rate
SL:	Arm-stroke length
BL-sLT:	Blood lactate concentration corresponding to the second lactate threshold
VO_2-sLT:	Oxygen uptake corresponding to the second lactate threshold
HR-sLT:	Heart rate corresponding to the second lactate threshold
SR-sLT:	Arm-stroke rate corresponding to the second lactate threshold
SL-sLT:	Arm-stroke length corresponding to the second lactate threshold
BL-T30:	Blood lactate concentration during 30 min of continuous swimming
VO_2-T30:	Oxygen uptake during 30 min of continuous swimming
HR-T30:	Heart rate during 30 min of continuous swimming
SR-T30:	Arm-stroke rate during 30 min of continuous swimming
SL-T30:	Arm-stroke length during 30 min of continuous swimming

References

1. Pyne, D.B.; Hamilton, L.; Swanwick, K.M. Monitoring the lactate threshold in world-ranked swimmers. *Med. Sci. Sports Exerc.* **2001**, *33*, 291–297. [CrossRef] [PubMed]
2. Zacca, R.; Azevedo, R.; Peterson Silveira, R.; Vilas-Boas, J.P.; Pyne, D.B.; Castro, F.A.d.S.; Fernandes, R.J. Comparison of incremental intermittent and time trial test in age group swimmers. *J. Strength Cond. Res.* **2019**, *33*, 801–810. [CrossRef] [PubMed]
3. Carvalho, D.C.; Soares, S.S.; Zacca, R.; Sousa, J.; Marinho, A.D.; Silva, J.A.; VilasBoas, J.P.; Fernandes, R.J. Anaerobic threshold biophysical characterization of the four swimming techniques. *Int. J. Sports Med.* **2020**, in press.
4. Arsoniadis, G.G.; Botonis, P.G.; Nikitakis, S.I.; Kalokiris, D.; Toubekis, A.G. Effects of successive annual training on aerobic endurance indices in young swimmers. *Open Sport Sci. J.* **2017**, *10*, 214–221. [CrossRef]
5. Fernandes, R.J.; Keskinen, K.; Colaço, P.; Querido, A.; Machado, L.; Morais, P.; Novais, D.; Marinho, D.; Vilas-Boas, J.P. Time limit at VO2max velocity in elite crawl swimmers. *Int. J. Sports Med.* **2008**, *29*, 145–150. [CrossRef]
6. Machado, L.; Almeida, M.; Morais, P.; Fernandes, R.J.; Vilas-Boas, J.P. Assessing the individual anaerobic threshold: The mathematical model. *Port. J. Sport Sci.* **2006**, *6*, 142–144.
7. Fernandes, R.J.; Sousa, M.; Machado, L.; Vilas-Boas, J.P. Step length and individual anaerobic threshold assessment in swimming. *Int. J. Sports Med.* **2011**, *32*, 940–946. [CrossRef]
8. Faude, O.; Kindermann, W.; Meyer, T. Lactate threshold concepts. How valid are they? *Sports Med.* **2009**, *39*, 469–490. [CrossRef]
9. Jamnick, A.N.; Botella, J.; Botella, J.; Pyne, D.B.; Bishop, D.J. Manipulating graded exercise test variables affects the validity of the lactate threshold and VO_2peak. *PLoS ONE* **2018**, *13*, 1–21. [CrossRef]
10. Beneke, R.; Hütler, M.; Von Duvillard, S.P.; Sellens, M.; Leithäuser, R.M. Effect of test interruptions on blood lactate during constant workload testing. *Med. Sci. Sports Exerc.* **2003**, *35*, 1626–1630. [CrossRef]
11. Borszcz, K.; Tramontin, A.F.; Bossi, A.H.; Carminatti, L.J.; Costa, V.P. Functional threshold power in cyclists: Validity of the concept and physiological responses. *Int. J. Sports Med.* **2018**, *39*, 737–742. [PubMed]
12. Ribeiro, J.; Toubekis, A.G.; Figuiredo, P.; de Jesus, K.; Tousaint, H.M.; Alves, F.; Vilas-Boas, J.P.; Fernandes, R.J. Biophysical determinants of front-crawl swimming at moderate and severe intensities. *Int. J. Sports Physiol. Perf.* **2017**, *12*, 241–246. [CrossRef] [PubMed]
13. Barbosa, T.M.; Fernandes, R.J.; Keskinen, K.L.; Vilas-Boas, J.P. The influence of stroke mechanics into energy cost of elite swimmers. *Eur. J. Appl. Physiol.* **2008**, *103*, 139–149. [CrossRef] [PubMed]
14. Jürimäe, J.; Haljaste, K.; Cicchella, A.; Latt, E.; Purge, P.; Leppik, A.; Jürimäe, T. Analysis of swimming performance from physical, physiological, and biomechanical parameters in young swimmers. *Pediatr. Exerc. Sci.* **2007**, *19*, 70–81. [CrossRef]
15. Cohen, J. *Statistical Power Analysis for the Behavioral Sciences*, 2nd ed.; Lawrence Erlbaum Associates: Hillsdate, NJ, USA, 1988.
16. Faul, F.; Erdfelder, E.; Lang, A.G.; Bucher, A. G*Power 3: A flexible statistical power analysis program for social, behavioral, and biomedical sciences. *Behav. Res. Methods* **2007**, *39*, 175–191. [CrossRef]
17. Bland, J.M.; Altman, D.G. Statistical methods for assessing agreement between two methods of clinical measurement. *Lancet* **1986**, *327*, 307–310. [CrossRef]
18. Tokmakidis, S.P.; Léger, A.L.; Pilianidis, C.T. Failure to obtain a unique threshold on the blood lactate concentration curve during exercise. *Eur. J. Appl. Physiol. Occup. Physiol.* **1998**, *77*, 332–342. [CrossRef]
19. Papoti, M.; Vitório, R.; Araújo, G.G.; Da Silva, R.A.; Santhiago, V.; Martins, B.L.; Cunha, A.S.; Gobatto, A.C. Determination of force corresponding to maximal lactate steady state in tethered swimming. *Int. J. Exerc.* **2009**, *2*, 269–279.
20. Baron, B.; Dekerle, J.; Depretz, S.; Lefevre, T.; Pelayo, P. Self-selected speed and maximal lactate steady state speed in swimming. *J. Sports Med. Phys. Fit.* **2005**, *45*, 1–6.
21. Pelarigo, G.J.; Denadai, S.B.; Greco, C.C. Stroke phases responses around maximal lactate steady state in front crawl. *J. Sci. Med. Sport* **2011**, *14*, 168. [CrossRef]
22. Beneke, R.; Leithäuser, M.R.; Ochentel, O. Blood lactate diagnostics in exercise testing and training. *Int. J. Sports Physiol. Perf.* **2011**, *6*, 8–24. [CrossRef] [PubMed]

23. Seifert, L.; Komar, J.; Leprêtre, P.M.; Lemaitre, F.; Chavallard, F.; Alberty, M. Swim specialty affects energy cost and motor organization. *Int. J. Sports Med.* **2010**, *31*, 624–630. [CrossRef] [PubMed]
24. Skorski, S.; Faude, O.; Urhausen, A.; Kindermann, W.; Meyer, T. Intensity control in swim training by means of the individual anaerobic threshold. *J. Strength Cond. Res.* **2012**, *26*, 3304–3311. [CrossRef] [PubMed]
25. Nikitakis, I.S.; Paradisis, P.G.; Bogdanis, G.C.; Toubekis, A.G. Physiological responses of continuous and intermittent swimming at critical speed and maximum lactate steady state in children and adolescent swimmers. *Sports* **2019**, *7*, 25. [CrossRef]
26. Coyle, E.F.; Gonzalez-Alonso, J. Cardiovascular drift during prolonged exercise: New perspectives. *Exerc. Sports Sci. Rev.* **2001**, *29*, 88–92.
27. Pelarigo, J.G.; Machado, L.; Fernandes, R.J.; Greco, C.C.; Vilas-Boas, J.P. Oxygen uptake kinetics and energy system's contribution around maximal lactate steady state swimming intensity. *PLoS ONE* **2017**, *12*, e167263. [CrossRef]
28. Dekerle, J.; Brickley, G.; Alberty, M.; Pelayo, P. Characterising the slope of the distance—Time relationship in swimming. *J. Sci. Med. Sport* **2010**, *13*, 365–370. [CrossRef]
29. Barbosa, T.M.; Bragada, A.G.; Reis, M.V.; Marinho, A.D.; Carvalho, C.; Silva, A.J. Energetics and biomechanics as determining factors of swimming performance: Updating the state of the art. *J. Sci. Med. Sport.* **2010**, *13*, 262–269. [CrossRef]
30. Alberty, M.; Sidney, M.; Huot-Marchand, F.; Hespel, M.J.; Pelayo, P. Intracyclic velocity variations and arm coordination during exhaustive exercise in front crawl stroke. *Int. J. Sports Med.* **2005**, *26*, 471–475. [CrossRef]
31. Oliveira, M.F.M. Metabolic and Swim Technique Responses During the Exercise at the Speed Corresponding to Maximal Lactate Steady State Determined Continuously and Intermittently. Master's Thesis, São Paulo State University, Rio Claro, Brazil, 2008.
32. Dekerle, J.; Nesi, X.; Lefevre, T.; Depretz, S.; Sidney, M.; Huot Marchand, F.; Pelayo, P. Stroking parameters in front crawl swimming and maximal lactate steady state speed. *Int. J. Sports Med.* **2005**, *25*, 53–58. [CrossRef]
33. Pelarigo, J.G.; Fernandes, R.J.; Ribeiro, L.; Benedito, D.; Greco, C.C.; Vilas-Boas, J.P. Comparison of different methods for the swimming aerobic capacity evaluation. *J. Strength Cond. Res.* **2018**, *32*, 3542–3551. [CrossRef] [PubMed]

Article

Validating Physiological and Biomechanical Parameters during Intermittent Swimming at Speed Corresponding to Lactate Concentration of 4 mmol·L^{-1}

Gavriil G. Arsoniadis [1], Ioannis S. Nikitakis [1], Petros G. Botonis [1], Ioannis Malliaros [1] and Argyris G. Toubekis [1,2,*]

1 Division of Aquatic Sports, School of Physical Education and Sports Science, National and Kapodistrian University of Athens, Dafne, 17237 Athens, Greece; garsoniadis@phed.uoa.gr (G.G.A.); inikitak@phed.uoa.gr (I.S.N.); pboton@phed.uoa.gr (P.G.B.); gmalliaros@phed.uoa.gr (I.M.)

2 Sports Performance Laboratory, School of Physical Education and Sport Science, National and Kapodistrian University of Athens, Dafne, 17237 Athens, Greece

* Correspondence: atoubekis@phed.uoa.gr; Tel.: +30-2107-2727-6049; Fax: +30-210-727-6139

Received: 22 January 2020; Accepted: 14 February 2020; Published: 18 February 2020

Abstract: Background: Physiological and biomechanical parameters obtained during testing need validation in a training setting. The purpose of this study was to compare parameters calculated by a 5 × 200-m test with those measured during an intermittent swimming training set performed at constant speed corresponding to blood lactate concentration of 4 mmol·L^{-1} (V4). Methods: Twelve competitive swimmers performed a 5 × 200-m progressively increasing speed front crawl test. Blood lactate concentration (BL) was measured after each 200 m and V4 was calculated by interpolation. Heart rate (HR), rating of perceived exertion (RPE), stroke rate (SR) and stroke length (SL) were determined during each 200 m. Subsequently, BL, HR, SR and SL corresponding to V4 were calculated. A week later, swimmers performed a 5 × 400-m training set at constant speed corresponding to V4 and BL-5×400, HR-5×400, RPE-5×400, SR-5×400, SL-5×400 were measured. Results: BL-5×400 and RPE-5×400 were similar ($p > 0.05$), while HR-5×400 and SR-5×400 were increased and SL-5×400 was decreased compared to values calculated by the 5 × 200-m test ($p < 0.05$). Conclusion: An intermittent progressively increasing speed swimming test provides physiological information with large interindividual variability. It seems that swimmers adjust their biomechanical parameters to maintain constant speed in an aerobic endurance training set of 5 × 400-m at intensity corresponding to 4 mmol·L^{-1}.

Keywords: intermittent swimming; swimming training; arm stroke rate; arm stroke length; validity

1. Introduction

Swimming performance during training or a year-round training plan is dependent on several interrelated changes of physiological and biomechanical parameters [1,2]. Coaches aim to optimize swimmer's training, facilitating performance improvement through the assessment of aerobic endurance and biomechanical parameters. Assessment of aerobic endurance parameters, such as speed corresponding to blood lactate concentration of 4 mmol·L^{-1} (V4) or speed corresponding to lactate threshold and biomechanical parameters, requires testing with progressively increasing swimming speed protocols [3,4].

V4 has been highlighted as one of the most commonly used indices for assessing swimming endurance [5,6]. It is suggested that V4 corresponds to lactate threshold or the onset of blood lactate accumulation and may be used for the adjustment of training pace during training for improvement of aerobic capacity [4,7,8]. Additionally, biomechanical parameters corresponding to V4 may be connected to performance changes [9]. Calculation of V4 requires drawing of a speed vs. blood lactate concentration curve after the completion of several repetitions of 200-m to 400-m swimming with progressively increasing speed [10,11]. However, various progressively increasing speed tests have been used since 1980s. Specifically, progressively increasing speed tests conducted with five repetitions of 4-min, 6-min or 8-min duration in running [5,6], four to five repetitions of 200-m [8,9], seven repetitions of 200-m [3,10] or seven repetitions of 300-m and 400-m front crawl swimming [9] have been used for testing.

Despite the various progressively increasing swimming speed tests used, a criticism of the validity of using a fixed lactate concentration of 4 mmol·L^{-1} as the concentration corresponding to lactate threshold has been raised [12] because of the high individual variability noticed on lactate concentration corresponding to the lactate threshold (i.e., 2–4 mmol·L^{-1} [13,14]). Whatever the case, testing the validity of predicting lactate concentration during a training set may be helpful for coaches since lactate concentration during endurance training sets may range from 2–6 mmol·L^{-1} [12]. Furthermore, it is expected that physiological and biomechanical variables calculated by a progressively increasing speed test will be reproduced in a training set applied a few days later. However, calculation of V4 and all subsequent physiological (i.e., heart rate, lactate) and biomechanical parameters (i.e., stroke rate and stroke length) rely on mathematical calculations and linear or nonlinear correlations which all present a measurement error [4]. In this case, the parameters calculated by the progressively increasing speed test may differ from those measured during a long-duration training set, and this has never been examined.

To our knowledge, there is no study comparing fixed lactate values predicted by the speed vs. lactate curve with those obtained during an intermittent constant speed training set in well-trained swimmers. The purpose of the present study was to examine the validity of physiological and biomechanical parameters during intermittent swimming at speed corresponding to blood lactate concentration of 4 mmol·L^{-1}.

2. Materials and Methods

2.1. Participants

Twelve regional- and national-level competitive male swimmers specializing in various competitive distances volunteered to participate in the study (see Table 1). Participants had competitive swimming training background of mean (standard deviation: SD) 8.5 (1.7) years, and they participated in a daily swimming training (6 days per week) with duration of approximately 2 h per session. Each participant provided written informed consent after a thorough explanation of the study. The local institutional review board approved the experimental procedures (Approval No. 1007/26-4-2017), which were in accordance with the Declaration of Helsinki for Human Subjects.

2.2. Study Design

Physiological and biomechanical parameters calculated by a 5 × 200-m progressively increasing speed swimming test were compared with those measured during a constant speed 5 × 400-m intermittent swimming training set in this study. The speed during the training set was prescribed based on the speed vs. lactate concentration curve drawn after a progressively increasing speed swimming test.

Table 1. Participant characteristics in the current study. The data are presented as mean values with SD in parentheses.

Variables	Swimmers (*n* = 12)
Age (y)	19.0 (2.2)
Body mass (kg)	74.4 (10.1)
Height (cm)	178.1 (7.9)
Fat mass (%)	13.0 (2.6)
Body Mass Index (%)	23.4 (1.5)
VO_2peak ($mL \cdot kg^{-1} \cdot min^{-1}$)	65.5 (11.4)
VO_2max ($mL \cdot kg^{-1} \cdot min^{-1}$)	51.2 (14.1)
Time (s), 200-m front crawl	127.1 (9.6)
FINA points, 200-m front crawl	534 (127)
Time (s), 400-m front crawl	288.7 (22.0)
FINA points, 400-m front crawl	501 (186)
FINA points of best competition	625 (149)

2.3. Preliminary Testing—The 400-m Test

The study was conducted during the specific preparation mesocycle of training and swimmers were tested in three testing sessions 48 h apart (Figure 1). All swimming tests were conducted in a 25-m indoor swimming pool with a constant temperature of 25–26 °C. During the first visit and following a standardized warm-up (400-m slow swimming at 60% intensity, 4 × 50-m front crawl leg kicking, 4 × 50-m front crawl drills and 4 × 50-m front crawl swim with progressively increasing speed), swimmers participated in a 400-m front crawl test with maximum intensity. Immediately after the completion of the 400-m test, a face mask was applied to the swimmer for expired gas collection during recovery and VO_2peak determination (VO2OOO; MedGraphics, Saint Paul, MN, USA; [15]).

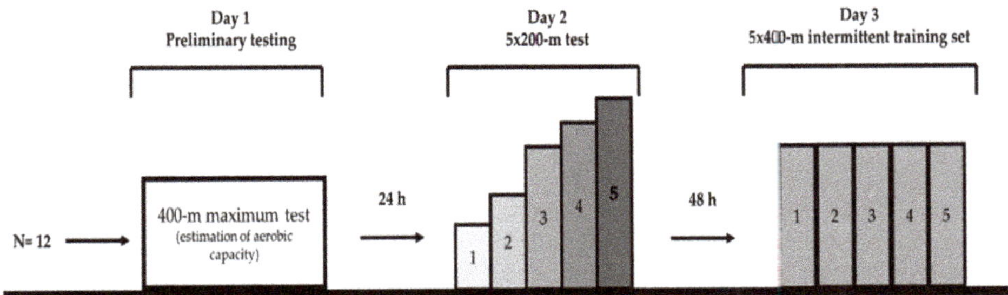

Figure 1. Study design of the current study; 400-m: 400-m front crawl; 5 × 200-m: five repetitions of 200-m front crawl, 5 × 400-m: five repetitions of 400-m front crawl, h: hours.

2.4. Applying the 5 × 200-m Test

On the following day, all swimmers participated in a standardized swimming warm-up (following the same procedure as before the 400-m test) and 10 minutes later performed a 5 × 200-m front crawl test at intensities corresponding to 60%, 70%, 80% and 90% of the 200-m maximum speed progressively during the first four repetitions, exerting maximum effort in the last 200-m repetition. During the 5 × 200-m test, each repetition started every 5.5 min with a push-off start from within the water. Fingertip blood samples were collected after each repetition and were analyzed for blood lactate concentration (BL) using the reflectance photometry enzymatic reaction method (Accutrend Plus; Roche, Germany). Rating of perceived exertion (RPE) was indicated using a 10-point scale after each 200-m repetition [16]. Heart rate (HR) was recorded continuously using telemetry (s610i; Polar Electro, Oy, Kempele, Finland). V4 was determined for each swimmer by interpolation from a second-order polynomial function of swimming speed vs. lactate concentration data (mean (SD) R^2 =

0.97 (0.03), mean r = 0.98 (0.01)). Heart rate corresponding to V4 (HR-V4) was determined for each swimmer individually by the linear regression between swimming speed and HR obtained during the 5 × 200-m test (mean R^2 = 0.96 (0.05), mean r = 0.98 (0.02)). Stroke rate (SR) was calculated by the time (T) to complete three arm-stroke cycles ($180 \cdot T^{-1}$), and stroke length (SL) was calculated by dividing swimming speed every 50 m (V) by SR. SR and SL corresponding to V4 (SR-V4 and SL-V4) were calculated by the interpolation of the best-fit regression line of SR and SL vs. swimming speed during the 5 × 200-m test (SR: mean R^2 = 0.97 (0.03), mean r = 0.99 (0.01); SL: mean R^2 = 0.98 (0.02), mean r = 0.99 (0.01)). Similarly, the rating of perceived exertion (RPE) corresponding to V4 was calculated by interpolation (mean R^2 = 0.97 (0.02), mean r = 0.98 (0.01)).

2.5. Intermittent Swimming Training Set of 5 × 400-m

Swimmers completed an intermittent swimming training set 48 h after the completion of the 5 × 200-m test. A standardized warm-up as described for the previous testing sessions was applied before the training set. Ten minutes after warm-up, swimmers completed a 5 × 400-m training set at a constant speed corresponding to V4 with a resting interval of 30 to 45 s between repetitions to allow blood sampling. Swimming speed was kept constant by using a sound transmitter attached next to the swimmer's ear (FINIS tempo pro, Finis Inc., Livermore, CA, USA) and according to the individual V4 that was determined by the 5 × 200-m test. Swimmers were advised to touch the swimming pool wall with their legs in each 25-m lap when hearing the transmitted sound. Additionally, one of the researchers recorded the time for each 50-m split in all 5 × 400-m repetitions (HS-80; CASIO, Guangzhou, China), and the mean speed of the test was calculated (V-5×400). BL concentration was collected after the first, third and fifth 400-m repetitions, while HR was recorded continuously and RPE was indicated after each repetition. SR and SL were calculated during each 400-m repetition of the 5 × 400-m swimming training set. The mean values of, BL-5×400, HR-5×400, RPE-5×400, SR-5×400 and SL-5×400 were used for the statistical analysis.

2.6. Statistical Analysis

Student's *t*-test for paired samples was used to compare physiological and biomechanical parameters corresponding to V4 and calculated after the 5 × 200-m test with those measured during the 5 × 400-m intermittent swimming training set with constant speed. Specifically, V4, BL-V4, HR-V4, RPE-V4, SR-V4 and SL-V4 were compared to V-5×400, BL-5×400, HR-5×400, RPE-5×400, SR-5×400 and SL-5×400. Pearson *r* correlations were used to examine the relationship between relevant parameters. Additionally, the effect size for paired comparisons was calculated with Cohen's *d* [17], using the pooled standard deviation as the denominator. The effect size was considered small if the absolute value of Cohen's *d* was less than 0.20, medium if it was between 0.20 and 0.50 and large if it was greater than 0.50. The 95% confidence limits (95% CL) were also calculated for the mean differences between parameters obtained by the two tests. Agreement of measured parameters was tested using Bland and Altman plots [18]. SPSS, software (v.23, SPSS Inc., Chicago, IL, USA) was used for data analysis. Data are presented as mean and SD (in parentheses). Statistical significance was set at $p < 0.05$.

3. Results

3.1. Swimming Speed Comparison between Tests

Swimming speed prescribed by the 5 × 200-m test (V4) was successfully reproduced during the 5 × 400-m training set (V4 = 1.325 (0.08) $m \cdot s^{-1}$; V-5×400-m = 1.287 (0.10) $m \cdot s^{-1}$; mean difference (SD): 0.04 (0.08) $m \cdot s^{-1}$; 95% CL: 0.060, 0.017 $m \cdot s^{-1}$; $d = -0.42$, $p = 0.122$). Additionally, significant correlation was noticed between V4 and V-5×400-m (r = 0.64, $p = 0.02$) and agreement was indicated by Bland and Altman plot (bias (SD): −0.04 (0.08) $m \cdot s^{-1}$).

3.2. Comparison of Physiological Variables and Rating of Perceived Exertion between Tests

Following the training set, BL-5×400 was not different compared to 4 mmol·L^{-1} (BL-V4 = 4.0 (0.0) mmol·L^{-1}; BL-5×400 = 5.0 (2.6) mmol·L^{-1}; mean difference (SD): −1.0 (2.6) mmol·L^{-1}; 95% CL: −1.53, −0.42 mmol·L^{-1}; d = 0.75, p = 0.231, Figure 2). Bland and Altman plot indicated agreement between calculated and measured values (bias (SD): −1.0 (2.6) mmol·L^{-1}). Training HR-5×400 was higher compared to HR-V4 (mean difference (SD): −7.2 (10.7) b·min^{-1}; 95% CL: −11.22, −3.11 b·min^{-1}; d = 0.59, p = 0.04, Figure 2) and these parameters were correlated (r = 0.63, p = 0.03). Bland and Altman plot showed agreement between calculated and measured values (bias (SD) 7 (11) b·min^{-1}).

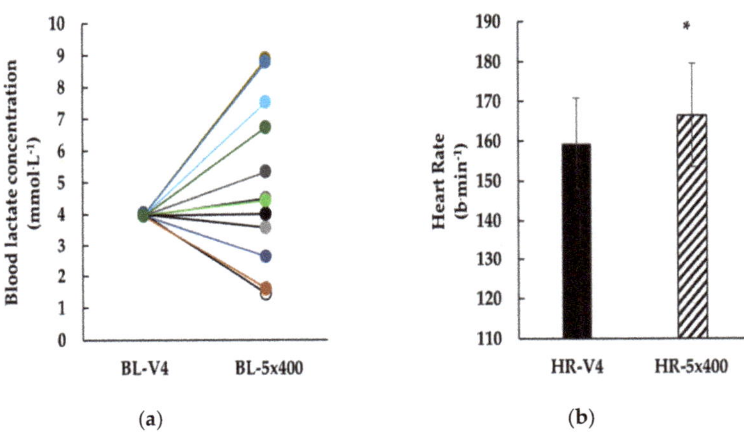

Figure 2. (**a**): Comparison between blood lactate concentration corresponding to V4 calculated after a progressively increasing speed 5 × 200-m test (BL-V4) and blood lactate measured during a constant speed intermittent swimming training set of 5 × 400-m (BL-5×400). (**b**): Comparison between heart rate corresponding to V4 calculated after a progressively increasing speed 5 × 200-m test (HR-V4) and heart rate measured during a constant speed intermittent swimming training set of 5 × 400-m (HR-5×400). *: $p < 0.05$ between HR-V4 and HR-5×400-m.

The calculated RPE values were successfully reproduced during the training set (RPE-V4 = 4.25 (2.09); RPE-5×400 = 5.17 (1.69); mean difference (SD): −1.0 (2.5); 95% CL: −1.6, −0.4; d = 0.52, p = 0.224]. Calculated and measured values were not correlated (r = 0.16, p = 0.60). Bland and Altman plot showed agreement between calculated and measured values (bias (SD): −1.0 (2.5)).

3.3. Comparison of Biomechanical Variables between Tests

Measured SR during the training set was increased compared to calculated SR and measured SL was decreased compared to calculated SL (Figure 3). Although a significant correlation was noticed between measured and calculated SR and SL values (SR, r = 0.57, p = 0.05; SL, r = 0.96, p = 0.001). Bland and Altman plot indicated agreement between calculated and measured for SR and SL (bias (SD): −5.6 (3.3) cycles·min^{-1} and 0.13 (0.09) m·cycles^{-1}; Figure 3).

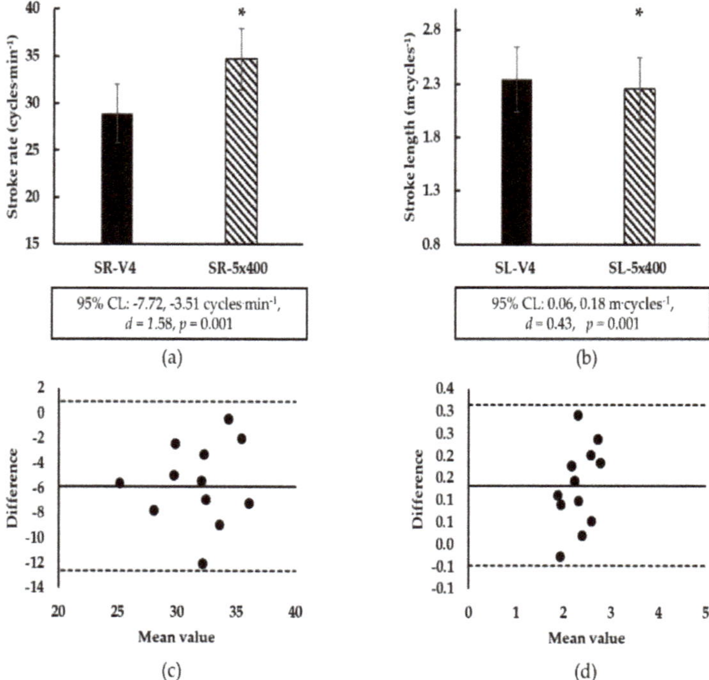

Figure 3. (**a**): Comparison of the stroke rate. (**b**): comparison of stroke length. Values corresponding to V4 and calculated after a progressively increasing speed 5 × 200-m test (SR-V4 and SL-V4) are compared with the stroke rate and stroke length measured during an intermittent constant swimming speed 5 × 400-m training set (SR-5×400 and SL-5×400). The 95% confidence limits (95% CL) and effect size (*d*) are indicated under the figures. (**c**): Bland and Altman plots of mean vs. difference in SR-V4 and SR-5×400 and (**d**): Bland and Altman plots of mean vs. difference in SL-V4 and SL-5×400. V4: speed corresponding to a blood lactate concentration of 4 mmol·L^{-1}. Units of measure in (c) and (d) are not shown for clarity and are the same as in the corresponding figure (**a**) and (**b**) panel. * $p < 0.05$ between SR-V4 and SR-5×400 and between SL-V4 and SL-5×400-m.

4. Discussion

The purpose of the current study was to validate the physiological and biomechanical parameters corresponding to a fixed lactate concentration of 4 mmol·L^{-1} with those measured in an intermittent 5 × 400-m swimming training set at constant speed. A constant swimming speed corresponding to V4 was successfully maintained and BL was no different from that at 4 mmol·L^{-1} during the 5 × 400-m training set. However, during the training set, a lower SL and a higher HR and SR compared to predicted values were recorded. Despite the Bland and Altman plot indicating agreement, a great individual variation was observed for all physiological and biomechanical variables.

A similar speed compared to V4 was expected during the 5 × 400-m training set as this was predesigned in our experimental design. This is because swimmers were guided to follow this speed using audio signals. Considering that swimmers maintained the prescribed speed, we should explain the observed metabolic and biomechanical alterations required by the swimmers to maintain this speed. No difference was found between 4 mmol·L^{-1} and BL-5×400-m, although a large effect size (d = 0.75) and a great variation in measured values was observed between swimmers. Using a fixed value of 4 mmol·L^{-1} has been criticized in previous studies since it does not taking into account an individual approach to estimate BL [3,12]. Specifically, it was proposed that the fixed value did not take into account the individual blood lactate concentration curve kinetics and can be affected by the muscle glycogen content [12,13]. This is confirmed by studies reporting a higher V4 or BL-V4 during a continuous swimming trial but not in an intermittent swimming set as in the current study [19,20]. However, it should be considered that in the aforementioned studies [19,20], well-trained long-distance swimmers or young swimmers were included. Well-trained swimmers specializing in various distances participated in the current study, and they may present different individual characteristics compared to swimmers participating in previous studies. The swimming stroke specialty or athlete specializations may differentiate BL efflux under a constant speed [20,21]. Specifically, it has been shown that sprint-oriented athletes or those with limited aerobic potential may present higher BL values compared with long distance athletes during a continuous swimming training set under a constant speed at V4 [21,22], as was observed in the current study. Despite the observed variation, BL concentration prescribed by the progressively increasing speed test was similar to that measured during the intermittent training set. Lactate responses are reflected by RPE, an index of internal training load, during intermittent swimming [23]; RPE calculated by the 5 × 200-m test was found to be similar to that measured during the 5 × 400-m training set.

Individual variation observed in measured BL and RPE, as well as differences in HR, may be attributed to equations used for calculation of these parameters. The second-order polynomial function used to calculate V4 presented a low error of estimation ($r = 0.98$, $SE = 0.07$) and a high accuracy of the predicted BL value. In contrast with BL, the measured HR values were higher during the 5 × 400-m intermittent training set compared to the value calculated by the 5 × 200-m progressively increasing speed test. This may be explained by a higher sympathetic activation during the 5 × 400-m intermittent training set [24]. Furthermore, the higher HR during the 5 × 400-m training set may indicate a higher effort of swimmers to maintain the prescribed V4 swimming speed. It should be noted that HR-V4 was calculated using an appropriate linear fit of data ($r = 0.98$) and low standard error of estimate ($3\ b{\cdot}min^{-1}$) that may partly explain the differences during 5 × 400-m training set and HR-V4. Possibly, differences in HR were dependent on exercise duration between progressively increasing speed testing protocol (200-m repetitions) and training set (400-m repetitions; [5]). In this case, HR may have not reached steady values within the shorter 200-m distance, thus underestimating the values during a longer duration 400-m distance.

Besides the physiological load required in maintaining constant V4, some differences in biomechanical parameters were also noticed. Increased SR-V4 (large effect size) and decreased SL-V4 (medium effect size) were observed during the 5 × 400-m intermittent training set compared to the calculated values by the progressively increasing speed 5 × 200-m test. Increased SR and decreased SL during swimming are connected to increased energy cost [25]. These changes occur when swimmers manage to maintain the prescribed speed. Possibly, well-trained swimmers who participated in the current study applied less force (decreased SL) to maintain an efficient arm stroke [26]. However, these changes were not severe enough to induce fatigue manifested as an inability to maintain constant speed during the 5 × 400-m. A further explanation for these differences may be that swimmers completed a higher number of arm-stroke cycles in the longer distance of 400-m compared to 200-m, thus altering their mechanics to compensate for the longer distance [27]. Possibly, the different distances used for testing (400-m vs. 200-m) as well as the swimmer's specialty combined with the large interindividual variation between swimmers may have led to the aforementioned differences in biomechanical parameters.

The characteristics of the progressively increasing speed testing protocols (number of repetitions, duration of each repetition) that have been used to estimate physiological and biomechanical variables may lead to different responses during an intermittent or a continuous training set [28,29]. However, this is controversial throughout the literature. Specifically, Madsen et al. [30], found that a progressively increasing speed test consisted of 200-m repetitions overestimates the physiological variables obtained during continuous swimming training, while recently it has been found that 200-m and 300-m testing protocols showed similar physiological and biomechanical variables during intermittent swimming training [31]. Whatever the case, coaches should be aware that the predicted parameters are dependent on testing protocol used for their calculation and may not be similar to those expected during an intermittent training set.

5. Conclusions

A 5 × 200-m progressively increasing speed test and the determination of V4 calculated by interpolation of a second-order polynomial function fitted to the swimming speed vs. lactate concentration data seem to provide physiological and biomechanical variables with a large interindividual variability. It should be expected that swimmers will adjust their mechanics at increased metabolic cost to maintain the required speed during an intermittent training set including 400-m repetitions. Specifically, increased SR and decreased SL may be observed when swimmers aim to maintain speed during a 5 × 400-m intermittent training set. It seems that swimmers change the arm stroke profile in aiming to maintain the required speed during a long-duration constant intensity set. The large interindividual variation between swimmers, possibly because of their specialty or the characteristics (i.e., number of repetitions, interval time, distance) of the test used to calculate the required parameters, should be considered. Coaches should be aware that prescribed physiological or mechanical parameters may be altered when swimmers follow a training pace corresponding to V4 during aerobic endurance training.

Author Contributions: Conceptualization, A.G.T. and G.G.A.; methodology, A.G.T., P.G.B., G.G.A.; software, G.G.A.; validation, A.G.T. and P.G.B.; formal analysis, P.G.B.; investigation, G.G.A., I.S.N. and I.M.; resources, A.G.T.; data curation G.G.A.; writing – original draft preparation, G.G.A.; writing—review and editing A.G.T.; visualization, P.G.B.; supervision, A.G.T.; project administration, A.G.T. All authors have read and agreed to the published version of the manuscript.

Funding: This research received no external funding.

Acknowledgments: The authors would like to thank the swimmers for their participation in the study.

Conflicts of Interest: The authors declare no conflict of interest, financial or otherwise.

References

1. Barbosa, T.M.; Bragada, A.G.; Reis, M.V.; Marinho, A.D.; Carvalho, C.; Silva, A.J. Energetics and biomechanics as determining factors of swimming performance: Updating the state of the art. *J. Sci. Med. Sport* **2010**, *13*, 262–269. [CrossRef] [PubMed]
2. Smith, D.J.; Norris, S.R.; Hogg, J.M. Performance evaluation of swimmers. Scientific tools. *Sports Med.* **2002**, *32*, 539–554. [CrossRef] [PubMed]
3. Pyne, D.B.; Hamilton, L.; Swanwick, K.M. Monitoring the lactate threshold in world-ranked swimmers. *Med. Sci. Sports Exerc.* **2001**, *33*, 291–297. [CrossRef] [PubMed]
4. Faude, O.; Kindermann, W.; Meyer, T. Lactate threshold concepts. How valid are they? *Sports Med.* **2009**, *39*, 469–490. [CrossRef] [PubMed]
5. Foxdal, P.; Sjödin, B.; Sjödin, A.; Östman, B. The validity and accuracy of blood lactate measurements for prediction of maximal endurance running capacity. Dependency of analyzed blood media in combination with different designs of the exercise test. *Int. J. Sports Med.* **1994**, *15*, 89–95. [CrossRef] [PubMed]
6. Allen, W.; Seals, D.R.; Hurley, B.; Ehsani, A.; Hagberg, J.M. Lactate threshold and distance running performance in young and older endurance athletes. *J. Appl. Physiol.* **1985**, *58*, 1281–1284. [CrossRef]
7. Mader, A. Evaluation of the endurance performance of marathon runners and theoretical analysis of test results. *J. Sports Med.* **1991**, *31*, 1–19.
8. Toubekis, A.G.; Tsami, A.P.; Smillios, I.G.; Douda, H.T.; Tokmakidis, S.P. Training induced changes on blood lactate profile and critical velocity in young swimmers. *J. Strength Cond. Res.* **2011**, *25*, 1563–1570. [CrossRef]
9. Anderson, M.; Hopkins, W.; Roberts, A.; Pyne, D. Ability of test measures to predict competitive performance in elite swimmers. *J. Sports Sci.* **2008**, *26*, 123–130. [CrossRef]

10. Toubekis, A.G.; Vasilaki, A.; Douda, H.; Gourgoulis, V.; Tokmakidis, S. Physiological responses during interval training at relative to critical velocity in young swimmers. *J. Sci. Med. Sport* **2011**, *14*, 363–368. [CrossRef]
11. Arsoniadis, G.G.; Botonis, P.G.; Nikitakis, S.I.; Kalokiris, D.; Toubekis, A.G. Effects of successive annual training on aerobic endurance indices in young swimmers. *Opens Sport Sci. J.* **2017**, *10*, 214–221. [CrossRef]
12. Fernandes, R.J.; Sousa, M.; Machado, L.; Vilas-Boas, J.P. Step length and individual anaerobic threshold assessment in swimming. *Int. J. Sports Med.* **2011**, *32*, 940–946. [CrossRef] [PubMed]
13. Reilly, T.; Woodbridge, V. Effects of moderate dietary manipulation on swim performance and on blood lactate swimming velocity curves. *Int. J. Sports Med.* **1999**, *20*, 93–97. [CrossRef] [PubMed]
14. Simon, G. The role of lactate testing in swimming. In Proceedings of the XII FINA World Congress on Sports Medicine, Goteborg, Sweden, 12–15 April 1997; pp. 259–262.
15. Costill, D.L.; Kovaleski, J.; Porter, D.; Fielding, R.; King, D. Energy expenditure during front crawl swimming: Predicting success in middle-distance events. *Int. J. Sports Med.* **1985**, *6*, 266–270. [CrossRef] [PubMed]
16. Borg, G. Perceived exertion as an indicator of somatic stress. *Scand. J. Rehabil. Med.* **1970**, *2*, 92–98. [PubMed]
17. Cohen, J. *Statistical Power Analysis for the Behavioral Sciences*, 2nd ed.; L. Erlbaum Associates: Hillsdate, NJ, USA, 1988.
18. Bland, J.M.; Altman, D.G. Statistical methods for assessing agreement between two methods of clinical measurement. *Lancet* **1986**, *327*, 307–310. [CrossRef]
19. Fernandes, R.J.; Keskinen, K.; Colaço, P.; Querido, A.; Machado, L.; Morais, P.; Novais, D.; Marinho, D.; Vilas-Boas, J.P. Time limit at V02max velocity in elite crawl swimmers. *Int. J. Sports Med.* **2008**, *29*, 145–150. [CrossRef]
20. Fernandes, R.J.; Sousa, M.; Pinheiro, A.; Vilar, S.; Colaço, P.; Vilas-Boas, J.P. Assessment of individual anaerobic threshold and stroking parameters in 10-11 years old swimmers. *Eur. J Sport Sci.* **2010**, *10*, 311–317. [CrossRef]
21. Stegmann, H.; Kindermann, W. Comparison of prolonged exercise tests at the individual anaerobic threshold and the fixed anaerobic threshold of 4 mmol·L^{-1} lactate. *Int. J. Sports Med.* **1982**, *3*, 105–110. [CrossRef]
22. Stegmann, H.; Kindermann, W.; Schnabel, A. Lactate kinetics and individual anaerobic threshold. *Int. J. Sports Med.* **1981**, *2*, 160–165. [CrossRef]
23. Wallace, L.; Slattery, K.; Couts, A. The ecological validity and application of the session—RPE method for quantifying training loads in swimming. *J. Strength Cond. Res.* **2009**, *23*, 33–38. [CrossRef] [PubMed]
24. Coyle, E.F.; Gonzalez-Alonso, J. Cardiovascular drift during prolonged exercise: New perspectives. *Exerc. Sport Sci. Rev.* **2001**, *29*, 88–92. [PubMed]
25. Barbosa, T.M.; Fernandes, R.J.; Keskinen, K.L.; Vilas-Boas, J.P. The influence of stroke mechanics into energy cost of elite swimmers. *Eur. J. Appl. Physiol.* **2008**, *103*, 139–149. [CrossRef]
26. Alberty, M.; Sidney, M.; Pelayo, P.; Toussaint, H.M. Stroking characteristics during time to exhaustion tests. *Med. Sci. Sports Exerc.* **2009**, *41*, 637–644. [CrossRef] [PubMed]
27. Zacca, R.; Azevedo, R.; Peterson Silveira, R.; Vilas-Boas, J.P.; Pyne, D.B.; Castro, F.A.d.S.; Fernandes, R.J. Comparison of incremental intermittent and time trial test in age group swimmers. *J. Strength Cond. Res.* **2019**, *33*, 801–810. [CrossRef]
28. Bentlay, J.D.; Newell, J.; Bishop, D. Incremental exercise test design and analysis. Implications for performance diagnostics in endurance athletes. *Sports Med.* **2007**, *37*, 575–586. [CrossRef]
29. Beneke, R.; Hütler, M.; Von Duvillard, S.P.; Sellens, M.; Leithäuser, R.M. Effect of test interruptions on blood lactate during constant workload testing. *Med. Sci. Sports Exerc.* **2003**, *35*, 1626–1630. [CrossRef]
30. Madsen, O.; Lohberg, M. The lowdown on lactate. *Swim. Tech.* **1987**, *24*, 21–26.
31. Ribeiro, P.F.L.; Lima, S.C.M.; Gobatto, A.C. Changes in physiological and stroking parameters during interval swims at the slope of the d-t relationship. *J. Sci. Med. Sport* **2010**, *13*, 141–145. [CrossRef]

Article

Physiological and Race Pace Characteristics of Medium and Low-Level Athens Marathon Runners

Aristides Myrkos [1], Ilias Smilios [1,*], Eleni Maria Kokkinou [1], Evangelos Rousopoulos [2*] and Helen Douda [1]

1 School of Physical Education and Sport Science, Democritus University of Thrace, 69100 Komotini, Greece; aris7tefaa@gmail.com (A.M.); ekokkino@phyed.duth.gr (E.M.K.); edouda@phyed.duth.gr (H.D.)

2 Ergoscan Physical Performance Evaluation Center, Ionias 110, 17456 Alimos, Greece; vrousso@gmail.com

* Correspondence: ismilios@phyed.duth.gr; Tel.: +30-2531039723

Received: 6 July 2020; Accepted: 19 August 2020; Published: 21 August 2020

Abstract: This study examined physiological and race pace characteristics of medium- (finish time < 240 min) and low-level (finish time > 240 min) recreational runners who participated in a challenging marathon route with rolling hills, the Athens Authentic Marathon. Fifteen athletes (age: 42 ± 7 years) performed an incremental test, three to nine days before the 2018 Athens Marathon, to determine maximal oxygen uptake (VO_2 max), maximal aerobic velocity (MAV), energy cost of running (ECr) and lactate threshold velocity (vLTh), and were analyzed for their pacing during the race. Moderate- (n = 8) compared with low-level (n = 7) runners had higher ($p < 0.05$) VO_2 max (55.6 ± 3.6 vs. 48.9 ± 4.8 $mL \cdot kg^{-1} \cdot min^{-1}$), MAV (16.5 ± 0.7 vs. 14.4 ± 1.2 $km \cdot h^{-1}$) and vLTh (11.6 ± 0.8 vs. 9.2 ± 0.7 $km \cdot h^{-1}$) and lower ECr at 10 km/h (1.137 ± 0.096 vs. 1.232 ± 0.068 $kcal \cdot kg^{-1} \cdot km^{-1}$). Medium-level runners ran the marathon at a higher percentage of vLTh (105.1 ± 4.7 vs. 93.8 ± 6.2%) and VO_2 max (79.7 ± 7.7 vs. 68.8 ± 5.7%). Low-level runners ran at a lower percentage ($p < 0.05$) of their vLTh in the 21.1–30 km (total ascent/decent: 122 m/5 m) and the 30–42.195 km (total ascent/decent: 32 m/155 m) splits. Moderate-level runners are less affected in their pacing than low-level runners during a marathon route with rolling hills. This could be due to superior physiological characteristics such as VO_2 max, ECr, vLTh and fractional utilization of VO_2 max. A marathon race pace strategy should be selected individually according to each athlete's level.

Keywords: endurance; aerobic performance; lactate threshold; running economy; maximal oxygen consumption; oxygen fractional utilization; running

---

1. Introduction

Marathon running is one of the most demanding races which requires well-organized mental and physical preparation [1]. Today, marathon races have turned into very large events where thousands of elite, high-level and recreational athletes participate in this 42.195 m race [2–4]. For many years, the physiological demands of a marathon as well as the physiological characteristics of top-class athletes were examined by researchers [5–11]. It is known that the most important parameters to sustain the highest possible running velocity over a marathon are the maximal oxygen uptake (VO_2 max), a high fractional utilization of VO_2 max and the energy cost of running (ECr) [7,12,13]. These parameters explain 70% of the variance of the average running speed sustained during a marathon race [6,7] and are good indicators of the endurance performance of individuals of different ages, genders and disciplines [1]. A typical VO_2 max value for male top-class marathoners is about 70–85 mL/kg/min, for low-level athletes around 65 mL/kg/min and for recreational runners about 51–58 mL/kg/min [14–16]. Additionally, oxygen fractional utilization at lactate threshold (LTh) intensity the point where blood lactate concentrations increase from baseline, is higher for top-class marathoners compared with low-level athletes (65–80% vs. 50–80% of VO_2 max, respectively) and is also higher at the lactate turn-point (LTP), the point where an abrupt increase in blood lactate is observed (85–90% vs. 80–85% of VO_2 max, respectively) [6,11,12,17,18].

Few studies have examined in more detail the physiological characteristics of recreational marathon runners, with finishing times >3 h, and how these characteristics affect performance in this group of runners. It was shown that the better the level of recreational marathoners, the higher the VO_2 max as well as the velocity and the VO_2 at LTh [2,19]. No differences were observed between the different level of runners in the LTh expressed as a percentage of VO_2 max and the oxygen cost of running at LTh [2]. Regarding medium- and low-level recreational runners, however, no data exist about the correspondence of race pace on the blood lactate curve, the fractional utilization of VO_2 max at race pace and if these differ according to the performance ability of the runners.

Most of the studies examined runners who participated in a marathon ran on a flat terrain where they could sustain a relatively stable pace till the end of the race, although the lower the level of the runners, the higher the variability in race pace [20]. The peculiarity of the terrain could be an external factor that may affect the physiological and race pace characteristics of a marathon race. The terrain at one of the most famous and challenging marathons in the world, the Athens Authentic Marathon, is characterized by rolling hills and includes the toughest uphill climb of any major marathon. The total ascent is 317 m (51.2% of the route is uphill), the total descent is 262 m (40.5% of the route is downhill) and the steepest grading ranges from −6.2 to 3.8% [21,22]. It is possible that the difficulty of the route may affect differently the race pace characteristics of medium- and low-level recreational runners. A runner with a faster pace will cross the hill segments in a shorter amount of time compared with a slower runner, altering probably the physiological requirements of the run. Therefore, recreational runners of different levels may run the Athens Marathon at a rate corresponding to different percentages of aerobic performance parameters. This may lead athletes and coaches to over- or underestimate the potential performance and to the determination of a false race pace strategy. Therefore, it would be useful to examine which are the physiological and race pace characteristics of medium- and low-level recreational athletes participating in the Athens Marathon and if they adopt different pace characteristics in relation to their physiological profile. Based on the above, the aim of the present study was to compare the physiological and race pace characteristics of medium- (finish time < 240 min) and low-level (finish time > 240 min) recreational runners who participated in the Athens Marathon.

2. Materials and Methods

2.1. Participants

Fifteen recreational marathon runners (age: 42 ± 7 years, height: 174.9 ± 6.5 cm and body mass: 72.8 ± 6.9 kg) volunteered to take part in the study. All participants were healthy and ran approximately 1–2 years on a systematic basis with a structured program with an average weekly load of 50–60 km. Based on their finishing time at the Athens Marathon 2018, athletes were divided into a moderate-level group, with finishing times < 240 min (n = 8), and a low-level group with finishing times > 240 min (n = 7). Before the start of the study, the institutional review board committee approved the experimental protocol in accordance with the Helsinki Declaration.

2.2. Maximal Incremental Test

Three to nine days before participation in the Athens Marathon 2018 race, participants performed a maximal incremental test on a treadmill (Technogym run race 1200, Italy) for the determination of VO_2 max, maximum aerobic velocity (MAV), maximum heart rate (HRmax), the relationship between blood lactate concentration and running velocity, oxygen consumption and running velocity, heart rate and running velocity, and the energy cost of running.

The protocol started at 7 km·h^{-1} and was increased by 1.5 km·h^{-1} every 3 min until volitional exhaustion. Treadmill grade was set at 1% throughout the protocol. Gas exchange was measured by the open circuit Douglas bag method as described by Cooke (2009). The subject breathed through a low-resistance 2-way Hans-Rudolph 2700 B valve (Shawnee, OK, USA). The concentrations of CO_2 and O_2 in the expired air were measured by using the Hi-tech (GIR 250) combined Oxygen and Carbon Dioxide Analyzer. The gas analyzers were calibrated continuously against standardized gases (15.35% O_2, 5.08% CO_2 and 100% N_2). Expired volume was measured by means of a dry gas meter (Harvard) previously calibrated against standard air flow with a 3 L syringe. Barometric pressure and gas temperature were recorded and respiratory gas exchange data for each work load (i.e., VO_2, VCO_2 and V_E) were determined based on the computations described by Cooke [23] when $V_E$atps, $FECO_2$ and FEO_2 are known. The highest VO_2 value obtained during a 30-sec time period during the incremental exercise test was recorded as the subject's $\dot{V}O_2$max. HR was continuously measured telemetrically (Polar RS400) and the highest 10 sec value was regarded as maximal. The test was considered as maximal when at least 3 of the following criteria were achieved: (a) visual exhaustion of the participants, (b) a plateau in oxygen consumption (<2 mL $kg^{-1}{\cdot}min^{-1}$) despite an increase in running velocity, (c) maximal HR higher than 90% of the predicted maximum (220-age) and (d) maximum respiratory exchange ratio > 1.1. MAV was calculated using the following formula: MAV ($km{\cdot}h^{-1}$) = Velocity of the last completed stage + (seconds run at last stage/180).

2.3. LTh and LTP Determination

At the end of each stage during the incremental test, approximately 0.3 μL of whole blood was collected from the fingertip and immediately analyzed for lactate concentration with a portable analyzer (Lactate Pro 2, Arkray Factory Inc., Koka-Shi, Japan) using an enzymatic-amperometric method. The individual relationships between blood lactate concentrations and running velocities were determined using an exponential model: $y = a + b \times \exp(x/c)$, where y = lactate concentration, x = running velocity and a, b and c are constants. The LTh and the LTP were identified as the velocities ($km{\cdot}h^{-1}$) at which blood lactate concentrations were increased by 0.3 and 1.5 $mmol{\cdot}L^{-1}$ from baseline values, respectively. Furthermore, LTh and LTP were expressed relative to MAV (%MAV) units, and based on the relationship between VO_2 and running velocity were also expressed in absolute ($mL{\cdot}kg^{-1}{\cdot}min^{-1}$) and relative (%$VO_2$ max) VO_2 max values.

2.4. Energy Cost of Running

The gas exchange data (VO_2, VCO_2) collected during the final 30 s of every 3-min stage up to the previous stage from the LTP were used for the calculations of the caloric cost of running. Substrate oxidation rate ($g{\cdot}min^{-1}$) was estimated using nonprotein respiratory quotient equations [24]:

Fat oxidation ($g{\cdot}min^{-1}$) = 1.6947 × VO_2 ($L{\cdot}min^{-1}$) − 1.7012 × VCO_2 ($L{\cdot}min^{-1}$)

Carbohydrate oxidation ($g{\cdot}min^{-1}$) = 4.5851 × VCO_2 ($L{\cdot}min^{-1}$) − 3.22259 × VO_2 ($L{\cdot}min^{-1}$)

The energy produced from each substrate was calculated by assuming an energy equivalent for 1 g of fat and carbohydrate of 9.75 and 4.07 kcal, respectively [25]. Total ECr was quantified from the sum of these values and was expressed in $kcal{\cdot}kg^{-1}{\cdot}km^{-1}$. The energy cost of running at 10 km/h and at the velocities corresponding to LTh and marathon race pace were estimated from the relationship between exergy cost and running velocity derived from the incremental test.

All the physiological data were analyzed after the completion of the marathon race to avoid any pacing strategies from the participants and their coaches based on the results of testing.

2.5. Route Characteristics and Race Pace Analysis

The profile of the Athens Marathon route includes rolling uphills and downhills. More specifically, when calculated in 450 m intervals, the total ascent, total descent, the percent of uphill distance, the percent of downhill distance and the steepest uphill and downhill, respectively, are for: (a) the total route: 317 m, 262 m, 51.2%, 40.5%, 3.8% and −6.2%, (b) the 0–10 km split: 19 m, 36 m, 36.4%, 50%, 1.3% and −2.0%, (c) the 10–21.1 km split: 143 m, 66 m, 66.7%, 25%, 3.6% and −6.2%, (d) the 21.1–30 km split:

122 m, 5 m, 95%, 5%, 3.3% and −1.1% and (e) the 30–42.195 km split: 32 m, 155 m, 15.4%, 76.9%, 3.3% and −5.1% [22].

Finishing time and split times for each participant were exported from the official results posted on the site of the organization [21]. The average running velocity of each runner was calculated by dividing marathon distance to the time needed to complete the race. Race pace was expressed as a percentage of VO_2 max (index of fractional utilization of VO_2 max), MAV and the velocities at LTh (vLTh) and LTP (vLTP).

To determine the differences between the two groups in pacing during the race, average running velocities for the distances of 0–10, 10–21.1, 21.1–30 and 30–42.195 km were calculated by dividing the distance of the split to the time to complete the split. For the analysis of the data, mean velocity of each split was expressed as a percentage of the vLTh. The velocity at LTh was selected as a reference point because for the whole sample, average marathon running velocity was equal to vLTh.

2.6. Statistical Analysis

All data are presented as means ± SD. Normality of the distribution of the data was examined with the Shapiro–Wilk's W test. A *t*-test was used to examine the differences among the medium-level and the low-level runners in the physiological parameters and race pace characteristics measured. A two-way analysis of variance with repeated measures in the second factor was used to examine the differences between the two groups in the mean velocity of each running split (0–10, 10–21.1, 21.1–30 and 30–42.195 km). Significant differences between means were located with the Newman–Keuls post hoc test. Pearson product moment correlations were used to determine the association between marathon time and the measured parameters. The statistical significance level was set for all tests at $p < 0.05$.

3. Results

3.1. Physiological Characteristics

Medium-level runners had higher ($p < 0.05$) VO_2 max, MAV, LTh ($km \cdot h^{-1}$), LTh (%MAV), LTh ($mL \cdot kg^{-1} \cdot min^{-1}$), LTP ($km \cdot h^{-1}$), LTP (%MAV) and LTP ($mL \cdot kg^{-1} \cdot min^{-1}$) than the low-level group. There were no significant differences ($p > 0.05$) between groups at HR_{max}, LTh (%VO_2 max) and LTP (%VO_2 max) (Table 1). Medium-level runners had lower ECr at 10 $km \cdot h^{-1}$ ($p = 0.05$), at vLTh ($p = 0.07$) and at marathon race pace ($p = 0.09$) (Table 1).

Table 1. Physiological and race pace characteristics of the medium-level, the low-level and of all runners.

	Medium-Level Runners	Low-Level Runners	All Runners	*p* Value between Groups
Age (years)	41.00 ± 7.69	42.14 ± 7.20	41.53 ± 7.22	0.36
Body height (m)	175.00 ± 6.44	174.71 ± 7.06	174.87 ± 6.49	0.94
Body mass (kg)	72.50 ± 6.58	73.17 ± 7.91	72.81 ± 6.97	0.86
VO_2 max ($mL \cdot kg^{-1} \cdot min^{-1}$)	55.56 ± 3.62	48.85 ± 4.77	52.43 ± 5.32	0.01
MAV ($km \cdot h^{-1}$)	16.45 ± 0.74	14.39 ± 1.24	15.49 ± 1.44	0.01
HRmax ($b \cdot min^{-1}$)	178.25 ± 9.54	183.71 ± 9.25	180.80 ± 9.5	0.28
vLTh ($km \cdot h^{-1}$)	11.58 ± 0.81	9.22 ± 0.72	10.48 ± 1.42	0.01
vLTP ($km \cdot h^{-1}$)	13.6 ± 0.87	11.1 ± 0.8	12.43 ± 1.52	0.01
vLT1% (%MAV)	70.37 ± 3.9	64.25 ± 4.15	67.51 ± 5.00	0.01
vLT2% (%MAV)	82.7 ± 4.83	77.37 ± 4.52	80.21 ± 5.29	0.05

Table 1. *Cont.*

	Medium-Level Runners	Low-Level Runners	All Runners	p Value between Groups
VO_2 LTh ($mL \cdot kg^{-1} \cdot min^{-1}$)	41.76 ± 1.81	35.11 ± 2.80	38.65 ± 4.10	0.01
VO_2 LTP ($mL \cdot kg^{-1} \cdot min^{-1}$)	48.04 ± 2.41	40.80 ± 3.78	44.66 ± 4.80	0.01
%VO_2 LTh (%VO_2 max)	75.31 ± 3.64	72.14 ± 5.87	73.83 ± 4.91	0.22
%VO_2 LTP (%VO_2 max)	86.59 ± 3.90	83.63 ± 4.08	85.21 ± 4.13	0.17
ECr 10 $km \cdot h^{-1}$ ($kcal \cdot kg^{-1} \cdot km^{-1}$)	1.137 ± 0.096	1.232 ± 0.068	1.181 ± 0.095	0.05
ECr vLTh ($kcal \cdot kg^{-1} \cdot km^{-1}$)	1.157 ± 0.079	1.232 ± 0.066	1.192 ± 0.081	0.07
ECr Race Pace ($kcal \cdot kg^{-1} \cdot km^{-1}$)	1.160 ± 0.083	1.232 ± 0.068	1.194 ± 0.082	0.09
Race pace ($km \cdot h^{-1}$)	12.14 ± 0.60	8.63 ± 0.64	10.50 ± 1.91	0.01
Race Pace (%MAV)	73.82 ± 2.60	60.11 ± 3.13	67.42 ± 7.59	0.01
Race Pace (%vLTh)	105.08 ± 4.71	93.80 ± 6.20	99.82 ± 7.84	0.01
Race Pace (%vLTP)	89.45 ± 4.47	77.92 ± 6.14	84.07 ± 7.85	0.01
Race Pace (%HRmax)	83.91 ± 5.8	77.41 ± 5.40	80.87 ± 6.37	0.04
Race Pace (%VO_2 max)	79.74 ± 7.65	68.80 ± 5.73	74.63 ± 8.68	0.01

HRmax: maximum heart rate, MAV: maximal aerobic velocity, vLTh: velocity ($km \cdot h^{-1}$) at lactate threshold, vLTP: velocity ($km \cdot h^{-1}$) at lactate turn-point, ECr: energy cost of running.

3.2. Race Pace Characteristics

Medium-level runners had, by design, a lower ($p < 0.05$) marathon time (209.0 ± 10.4 min, range: 194–225 min) than the low-level runners (289.7 ± 25.1 min, range: 260–328 min). Marathon finishing time was not related to the number of days between the maximal incremental test and the race day (Figure 1). Medium-level runners had a higher ($p < 0.05$) race pace expressed as %MAV, %vLTh, %vLTP, %VO_2 max and %HRmax (Table 1). Medium- and low-level runners had a similar ($p > 0.05$) race pace (expressed as %vLTh) at the first two running splits (0–10 and 10–21.1 km). However, low-level runners had a lower ($p < 0.05$) race pace at the last two splits (21.1–30 and 30–42.195 km) compared to the medium-level runners (Figure 2).

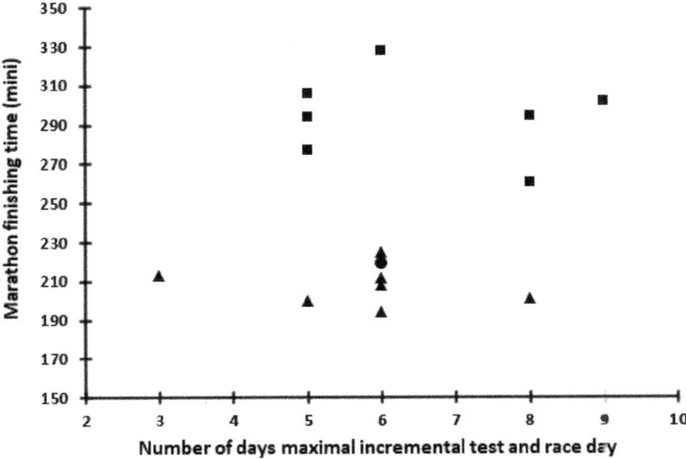

Figure 1. Plot of marathon finishing time vs. number of days between maximal incremental test and race day for the low-(squares) and the medium-level runners (triangles).

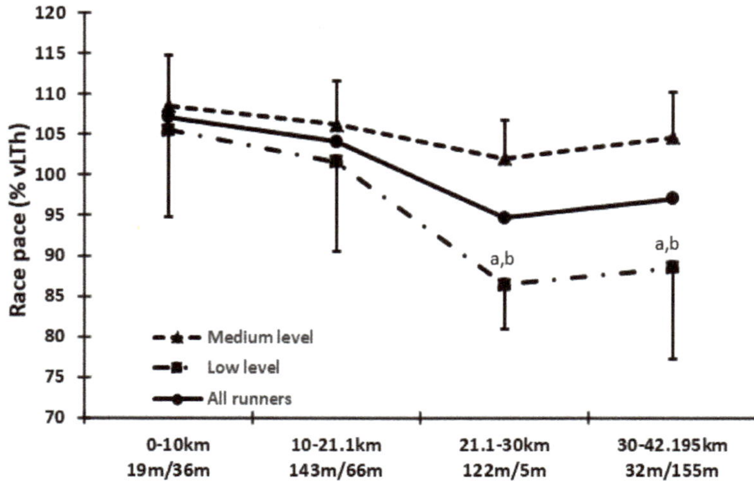

Figure 2. Race pace, expressed as a percentage of the velocity at lactate threshold (%vLTh), at the running splits of 0–10, 10–21.1, 21.1–30 and 30–42.195 km (total ascent in meters/total decent in meters) of the Athens Marathon, for the low-level, the medium-level and all runners. a: $p < 0.05$ significant difference between low- and medium-level runners, b: $p < 0.05$ significantly different from the 0–10 and 10–21.1 km splits for the low-level runners.

3.3. Correlation between Marathon Time and Measured Variables

Marathon finish time correlated significantly ($p < 0.05$) with VO_2 max (r = −0,76), MAV (r = −0.88), vLTh (km·h^{-1}; r = −0.91), vLTP (km·h^{-1}; r = −0.88), LTh (%MAV; r = −0.58), LTh (mL·kg^{-1}·min^{-1}; r = −0.86), LTP (mL·kg^{-1}·min^{-1}; r = −0.80), ECr 10 km·h^{-1} (r = 0.62), ECr vLTh (r = 0.59), ECr race pace (r = 0.55), race pace (%VO_2 max; r = −0.62), race pace (%vLTh; r = −0.75), race pace (%vLTP; r = −0.81) and race pace (%MAV; r = −0.90). Marathon finish time did not correlate significantly ($p > 0.05$) with LTP (%MAV; r = −0.38), LTh (%VO_2 max; r = −0.22) and LTP (%VO_2 max; r = −0.21).

4. Discussion

The purpose of this study was to provide further insight into the physiological and race pace characteristics of medium- and low-level marathon runners with a completion time < 240 min and > 240 min, respectively, of the Athens Authentic Marathon. This marathon race is famous, not only for historical reasons but also for its level of difficulty due to the peculiarity of the terrain. The results of the present study show that recreational medium-level runners compared to lower-level runners have: (a) higher VO_2 max, MAV and lactate threshold values in absolute velocity (km·h^{-1}) and VO_2 (mL·kg^{-1}·min^{-1}) units, (b) higher lactate threshold in relative velocity units (%MAV), (c) lower energy cost of running at 10 km/h and (d) adopt a race pace corresponding to a higher percentage of their lactate threshold velocity and fractional utilization of VO_2 max and show no significant alterations in their pace due to terrain alterations in contrast to the low-level runners to whom the uphill part of the race leads to great reductions in race pace.

Previous studies have examined the importance of physiological parameters and race pace characteristics of elite marathoners, but few studies provide data for recreational runners [2,19,20]. Maximal oxygen consumption, a high fractional utilization of VO_2 max and the energy cost of running are considered the determinants of endurance performance [26]. Indeed, in the present study the medium-level runners had higher VO_2 max than the low-level runners. This agrees with previous reports where the better level marathoners had higher VO_2 max than the lower level [2,7,18,19].

The VO_2 max values of the medium-level marathoners (55.56 ± 3.62 mL/kg/min) measured in the present study are approximately the same (55.7 ± 4.8) as those reported by Gordon et al. [2] for athletes who ran the marathon between 3:00 and 3:30 h as in the present study. The same holds even for the low-level runners (VO_2 max: 48.85 ± 4.77 mL/kg/min; finish time: 4:00–5:30 h) of the present study and runners with approximately similar finishing times in Gordon et al.'s [2] (VO_2 max: 46.5 ± 5.2 mL/kg/min; finish time: >4:30 h) and Chmura et al.'s [14] (VO_2 max: 51 ± 2 mL/kg/min; finish time: 4:17 ± 10.51 min) studies. It appears that certain levels of VO_2 max are necessary to achieve certain marathon times regardless of the level of the runner since VO_2 max determines the upper limit of aerobic performance. The high correlation between VO_2 max and marathon performance which has been previously reported for high-level to elite athletes [7,18,27,28] supports this notion. A large correlation (r = −0.76) between VO_2 max and marathon performance was observed as well in the present study for medium- to low-level marathon runners, enriching the limited information available for recreational athletes [2,19].

For most sports scientists, running economy or energy cost of running is a key factor for performance in long distance events and becomes more important as running distance increases [29–32]). In the present study, we examined ECr at a specific speed (10 km/h) and at the vLTh and we found that medium-level runners had lower ECr than the lower-level runners. This was probably another factor that allowed them to run the marathon at a faster pace. It should be noted that the ECr in the medium-level runners tended to be lower in the race pace as well. This is of importance considering that the medium-level runners sustained a faster running pace. Furthermore, ECr at 10 km/h and at vLTh had large correlations with marathon time (r = 0.62 and r = 0.59, respectively). The results of the current study reveal that running economy is a determinant of performance even for recreational runners with limited training experience and supports the suggestion in the literature that athletes should focus their training on the optimization of this parameter as well [32–34].

Besides the importance of VO_2 max and energy cost of running for marathon performance, stronger associations are observed between maximal aerobic velocity and the velocities at the lactate threshold or any point on the blood lactate curve, and endurance performance [35–37]. Similarly, very large correlations were observed in this study between marathon time and velocities at LTh and LTP (r = −0.91 and −0.88) and MAV (r = −0.88) of recreational runners. The velocity at LTh was the stronger single predictor of marathon finish time. This is not surprising considering that these indexes, when expressed in velocity units, encompass both VO_2 max and running economy [37]. When LTh and LTP values were normalized to MAV and VO_2 max, the relationship of these parameters with running performance became lower (r = −0.22 to −0.58). This is because the effect of VO_2 max and/or running economy was diminished [37]. Medium-level runners had higher LTh values, expressed either in velocity or VO_2 units, than the lower-level runners. Even when LTh velocity was normalized to MAV, medium-level runners had a higher LTh, indicating a higher ability of the fat oxidation rate to meet ATP demands and the occurrence of a later increased stimulation of glycolysis and glycogenolysis relative to their maximum performance. This probably reflects a greater aerobic capacity and increased buffering capacity promoting the ability to achieve higher running velocities due to metabolic and/or locomotor reasons.

Many studies declare that the fractional utilization of VO_2 max at LTh, LTP and at race pace is one of the most crucial parameters of aerobic performance along with VO_2 max and running economy [6,7,31]. Fractional utilization of VO_2 max at LTh and LTP did not differ between the two groups in the present study. Similarly, Gordon et al. [2] did not find any differences in fractional utilization of VO_2 max at LTh and LTP between recreational runners with different marathon finish times. It could be that adaptations in the utilization of oxygen from working muscles may require a significant amount of training load which was not achieved by our runners. In the present study, however, we found that medium-level runners had a higher fractional utilization of VO_2 max at marathon race pace. This agrees with previous findings that in high-level athletes, increased levels of fractional utilization of VO_2 max at marathon race pace were associated with faster performance [1,2,7].

In addition, a positive correlation of fractional utilization at race pace and marathon time was found (r = −0.62), Therefore, our data reveal that even in recreational runners, fractional utilization of VO_2 max at marathon race pace appears to be a contributing factor to performance.

A main finding of the present study is that medium-level marathoners ran the marathon distance with an average speed corresponding to higher percentages of vLTh and MAV. The better running economy may allow them to adopt a higher running velocity. Furthermore, the higher LTh and LTP velocities mean that the medium-level runners will cover a given distance at a shorter time. This may allow them to run at a higher point on the blood lactate curve because they can sustain this pace for the time needed to complete the race. On the contrary, the slower LTh and LTP velocities of the low-level runners mean that they need to run for a longer time to complete the race having a lower fractional oxygen utilization. It has been shown that as the duration of an endurance event increases, fractional oxygen utilization decreases [38,39]. The lower running velocity of the low-level runners made them spend more time running the uphill part of the Athens Marathon course. The total ascent from the 21.1st km to the 30th km is about 122 m and almost all this split is uphill. This forced the low-level runners to adopt an even lower velocity during this part of the route. Indeed, the split analysis revealed that the low-level runners were more influenced by this uphill part than the medium level. It appears that this specific segment has the greatest impact in the finish time between different levels of athletes and makes the Athens Marathon a totally different terrain from other marathons. It is worth noting that even at the last part of the route, which is mostly downhill, low-level runners were not able to increase their speed. Probably, the accumulated fatigue after hours of running may increase even more the stress placed on the musculoskeletal system, besides that induced by the increased eccentric load during downhill running, which prevents an increase in running speed compared to the previous uphill part. Therefore, the peculiarity of the terrain may affect differently the performance of a marathon runner depending on his/her ability level. This is of importance for coaches and athletes for the determination of the pace strategy to follow when running on a rolling hill terrain.

An advantage of the present study is that all recreational runners participated in the same marathon race and not in different ones. This makes comparisons between different levels of runners more reliable since all of them competed in the same route, on the same day and under the same environmental conditions. In addition, physiological testing was performed at a time point very close to the race day (three–nine days before) providing valid data about the relationship of physiological determinants of endurance performance and actual marathon running performance. Limitations of the present study, though, should also be acknowledged. A larger sample size would have given more valid data about the different levels of marathon runners. It was difficult, however, to measure many runners at a time close to the actual race. Furthermore, the energy cost of running at 10 km/h and at the velocity corresponding to LTh was estimated from the relationship between exergy cost and running velocity derived from the incremental test. Measurements at the exact velocities would have given more precise values of the energy cost. Again, the execution of these submaximal measurements would have increased the time of testing and it would not be possible to perform them near the race date.

5. Conclusions

The results of the present study enrich the existing literature regarding the physiological profile and the race pace characteristics of recreational marathon runners competing in a difficult route, the Athens Marathon. Medium-level runners (finish time range: 194–225 min) have higher VO_2 max, lactate threshold values, better running economy, greater oxygen fractional utilization at race pace and adopt a faster race pace in relation to their lactate threshold velocity than low-level runners (finish time range: 260–328 min). Furthermore, medium-level runners show no significant alterations in their pace due to terrain alterations in contrast to the low-level runners to whom the uphill part of the race leads to great reductions in race pace. Therefore, slower runners are more influenced by a hilly terrain and they decrease more their running velocity to complete this part of the race. Thus, careful planning of race pace should be considered so that pacing of the parts before the uphill would be of such an intensity to avoid a large decrease in running velocity at uphill. Therefore, besides the focus on training for the improvement of important physiological parameters related to endurance performance, it is recommended that the selected race pace strategy be applied individually according to each athlete's level.

Author Contributions: Conceptualization, A.M. and I.S.; methodology, A.M., I.S., E.M.K., E.R. and H.D.; investigation, E.M.K. and E.R.; data analysis, A.M., I.S., E.M.K. and E.R.; writing—original draft preparation, A.M., I.S. and H.D.; writing—review and editing, A.M., I.S. and H.D.; supervision, I.S. and H.D. All authors have read and agreed to the published version of the manuscript.

Funding: This research received no external funding.

Acknowledgments: The authors wish to thank the recreational runners who participated in the study.

Conflicts of Interest: The authors declare no conflict of interest.

References

1. Zinner, C.; Sperlich, B. *Marathon Running: Physiology, Psychology, Nutrition and Training Aspects*; Springer International Publishing: Berlin, Germany, 2016.
2. Gordon, D.; Wightman, S.; Basevitch, I.; Johnstone, J.; Espejo-Sanchez, C.; Beckford, C.; Boal, M.; Scruton, A.; Ferrandino, M.; Merzbach, V.; et al. Physiological and training characteristics of recreational marathon runners. *Open Access J. Sports Med.* **2007**, *8*, 231–241. [CrossRef] [PubMed]
3. Berndsen, J.; Smyth, B.; Lawlor, A. Pace my race: Recommendations for marathon running. In Proceedings of the 13th ACM Conference on Recommender Systems, Copenhagen, Denmark 16–20 September 2019.
4. Fokkema, T.; van Damme, A.; Fornerod, M.; de Vos, R.J.; Bierma-Zeinstra, S.; van Middelkoop, M. Training for a (half-)marathon: Training volume and longest endurance run related to performance and running injuries. *Scand. J. Med. Sci. Sports* in press. [CrossRef] [PubMed]
5. Foster, C.; Daniels, J.; Yarbrough, R. Physiological and training correlates of marathon running performance. *Aust. J. Sports Med.* **1977**, *9*, 58–61.
6. Di Prampero, P.E.; Atchou, G.; Brückner, J.-C.; Moia, C. The energetics of endurance running. *Eur. J. Appl. Physiol. Occup. Physiol.* **1986**, *55*, 259–266. [CrossRef]
7. Billat, V.; Demarle, A.; Slawinski, J.; Paiva, M.; Koralsztein, J.-P. Physical and training characteristics of top-class marathon runners. *Med. Sci. Sports Exerc.* **2001**, *33*, 2089–2097. [CrossRef]
8. Billat, V.; Demarle, A.; Paiva, M.; Koralsztein, J.P. Effect of Training on the Physiological Factors of Performance in Elite Marathon Runners (Males and Females). *Int. J. Sports Med.* **2002**, *23*, 336–341. [CrossRef]
9. Billat, V.; Petot, H.; Landrain, M.; Meilland, R.; Koralsztein, J.P.; Mille-Hamard, L. Cardiac Output and Performance during a Marathon Race in Middle-Aged Recreational Runners. *Sci. World J.* **2012**, *2012*, 1–9. [CrossRef]
10. Schmid, W.; Knechtle, B.; Knechtle, P.; Barandun, U.; Rüst, C.A.; Rosemann, T.; Lepers, R. Predictor Variables for Marathon Race Time in Recreational Female Runners. *Asian J. Sports Med.* **2012**, *3*, 90–98. [CrossRef]
11. Vernillo, G.; Schena, F.; Galvani, C.; Maggioni, M.A. A thropometric characteristics of top class Kenyan marathon runners. *J. Sports Med. Physic. Fit.* **2013**, *53*, 403–408.
12. Joyner, M.J.; Coyle, E.F. Endurance exercise performance: The physiology of champions. *J. Physiol.* **2008**, *586*, 35–44. [CrossRef]
13. Jones, A.M. The Physiology of the World Record Holder for the Women's Marathon. *Int. J. Sports Sci. Coach.* **2006**, *1*, 101–116. [CrossRef]
14. Chmura, J.; Chmura, P.; Konefat, M.; Batra, A.; Mroczek, D.; Kosowski, M.; Mlynarska, K.; Andrzejewski, D.; Rokita, A.; Ponikowski, P. The effects of a marathon effort on psychomotor performance ad catecholamine concentration in runners over 50 years of age. *Appl. Sci.* **2020**, *10*, 2067. [CrossRef]
15. Hernando, C.; Hernando, C.; Collado, E.J.; Panizo, N.; Martinez-Navarro, I.; Hernando, B. Establishing cut-points for physical activity classification using triaxial accelerometer in middle-aged recreational marathoners. *PLoS ONE* **2018**, *13*, e0202815. [CrossRef] [PubMed]
16. Hernando, C.; Hernando, C.; Martinez-Navaro, I.; Collado-Boira, E.; Panizo, N.; Hrenando, B.; Hernando, B. Estimation of energy consumed by midlle aged recreational marathoners during a marathon using accelerometry-based devices. *Sci. Rep.* **2020**, *10*, 1523. [CrossRef] [PubMed]

17. Beneke, R.; Leithäuser, R.M.; Ochentel, O. Blood Lactate Diagnostics in Exercise Testing and Training. *Int. J. Sports Physiol. Perform.* **2011**, *6*, 8–24. [CrossRef]
18. Sjödin, B.; Svedenhag, J. Applied Physiology of Marathon Running. *Sports Med.* **1985**, *2*, 83–99. [CrossRef]
19. Christensen, C.L.; Ruhling, R.O. Physical characteristics of novice and experienced women marathon runners. *Br. J. Sports Med.* **1983**, *17*, 166–171. [CrossRef]
20. Haney, T.A.; Mercer, J. A Description of Variability of Pacing in Marathon Distance Running. *Int. J. Exerc Sci.* **2011**, *4*, 133–140.
21. Available online: www.athensauthenticmarathon.gr (accessed on 2 August 2020).
22. Available online: www.plotaroute.com/routeprofile/20092 (accessed on 2 August 2020).
23. Cooke, C. Maximal oxygen uptake, economy and efficiency. In *Kinanthropometry and Exercise Physiology Laboratory Manual*, 3rd ed.; Eston, R., Reilly, T., Eds.; Taylor & Francis Group: London, UK, 2009; pp. 145–184.
24. Péronnet, F.; Massicotte, D. Table of nonprotein respiratory quotient: An update. *Can. J. Sport Sci.* **1991**, *16*, 23–29.
25. Jeukendrup, A.E.; Wallis, G.A. Measurement of Substrate Oxidation during Exercise by Means of Gas Exchange Measurements. *Int. J. Sports Med.* **2005**, *26*, S28–S37. [CrossRef]
26. Joyner, M.J.; Hunter, S.K.; Lucia, A.; Jones, A.M. Last Word on Viewpoint: Physiology and fast marathons. *J. Appl. Physiol.* **2020**, *128*, 1086–1087. [CrossRef] [PubMed]
27. Joyner, M.J. Modeling: Optimal marathon performance on the basis of physiological factors. *J. Appl. Physiol.* **1991**, *70*, 683–687. [CrossRef] [PubMed]
28. Lucia, A.; Esteve, J.; Olivan, J.; Gallego-Gomez, F. Physiological characteristics of the best Eritrean runners—Exceptional economy. *Appl. Physiol. Nutr. Met.* **2006**, *31*, 1–11. [CrossRef] [PubMed]
29. Foster, C.; Lucia, A. Running economy: The forgotten factor in elite performance. *Sports Med.* **2007**, *37*, 316–319. [CrossRef]
30. Saunders, P.; Pyne, D.; Telford, R.; Hawley, J. Factors affecting running economy in trained distance runners. *Sports Med.* **2004**, *34*, 465–485. [CrossRef]
31. Barnes, K.; Kilding, A. Running economy: Measurement, norms, and determining factors. *Sports Med. Open* **2015**, *1*, 8. [CrossRef]
32. Hoogkamer, W.; Kram, R.; Arellano, C.J. How Biomechanical Improvements in Running Economy Could Break the 2-h Marathon Barrier. *Sports Med.* **2017**, *47*, 1739–1750. [CrossRef]
33. Lee, E.; Snyder, E.; Lundstrom, C. Effects of marathon training in maximal aerobic capacity and running economy in experienced marathon runners. *J. Hum. Sport Exerc.* **2020**, *15*, 79–93.
34. Barnes, K.R.; Hopkins, W.G.; McGuigan, M.R.; Northuis, M.E.; Kilding, A.E. Effects of Resistance Training on Running Economy and Cross-country Performance. *Med. Sci. Sports Exerc.* **2013**, *45*, 2322–2331. [CrossRef]
35. Kolbe, T.; Dennis, S.; Selley, E.; Noakes, T.; Lambert, M. The relationship between critical power and running performance. *J. Sports Sci.* **1995**, *13*, 265–269. [CrossRef]
36. Noakes, T.D.; Myburgh, K.H.; Schall, R. Peak treadmill running velocity during theVO2max test predicts running performance. *J. Sports Sci.* **1990**, *8*, 35–45. [CrossRef] [PubMed]
37. Tokmakidis, S.P.; Léger, L.A.; Pilianidis, T.C. Failure to obtain a unique threshold on the blood lactate concentration curve during exercise. *Eur. J. Appl. Physiol. Occup. Physiol.* **1998**, *77*, 333–342. [CrossRef] [PubMed]
38. Davies, C.T.M.; Thompson, M.W. Aerobic performance of female marathon and male ultramarathon athletes. *Eur. J. Appl. Physiol. Occup. Physiol.* **1979**, *41*, 233–245. [CrossRef] [PubMed]
39. Maughan, R.J.; Leiper, J.B. Aerobic capacity and fractional utilization of aerobic capacity in elite and non-elite male and female marathon runners. *Eur. J. Appl. Physiol. Occup. Physiol.* **1983**, *52*, 80–87. [CrossRef] [PubMed]

Article

Acute Effects of Intermittent and Continuous Static Stretching on Hip Flexion Angle in Athletes with Varying Flexibility Training Background

Olyvia Donti *, Vasiliki Gaspari, Kostantina Papia, Ioli Panidi, Anastasia Donti and Gregory C. Bogdanis *

School of Physical Education & Sport Science, National and Kapodistrian University of Athens, 17237 Athens, Greece; vgaspari@phed.uoa.gr (V.G.); konpapia@phed.uoa.gr (K.P.); ipanidi@phed.uoa.gr (I.P.); adonti@phed.uoa.gr (A.D.)

* Correspondence: odonti@phed.uoa.gr (O.D.); gbogdanis@phed.uoa.gr (G.C.B.); Tel.: +30-21-0727-6109 (O.D.); +30-21-0727-6115 (G.C.B.)

Received: 16 February 2020; Accepted: 2 March 2020; Published: 3 March 2020

Abstract: This study examined changes in hip joint flexion angle after an intermittent or a continuous static stretching protocol of equal total duration. Twenty-seven female subjects aged 19.9 ± 3.0 years (14 artistic and rhythmic gymnasts and 13 team sports athletes), performed 3 min of intermittent (6 × 30 s with 30 s rest) or continuous static stretching (3 min) of the hip extensors, with an intensity of 80–90 on a 100-point visual analogue scale. The order of stretching was randomized and counterbalanced, and each subject performed both conditions. Hip flexion angle was measured with the straight leg raise test for both legs after warm-up and immediately after stretching. Both stretching types equally increased hip flexion angle by ~6% (continuous: 140.9° ± 20.4° to 148.6° ± 18.8°, $p = 0.047$; intermittent: 141.8° ± 20.3° to 150.0° ± 18.8°, $p = 0.029$) in artistic and rhythmic gymnasts. In contrast, in team sports athletes, only intermittent stretching increased hip flexion angle by 13% (from 91.0° ± 7.2° to 102.4° ± 14.5°, $p = 0.001$), while continuous stretching did not affect hip angle (from 92.4° ± 6.9° vs. 93.1° ± 9.2°, $p = 0.99$). The different effect of intermittent vs. continuous stretching on hip flexion between gymnasts and team sports athletes suggests that responses to static stretching are dependent on stretching mode and participants training experience.

Keywords: range of motion; hamstrings; stretching exercises; gymnastics; team sports; straight leg raise

---

1. Introduction

Warm-up routines usually include static stretching in its simplest form [1]. Static stretching is a widely used type of flexibility training [2,3] which is used to increase joint range of motion (ROM) in a duration dependent manner [4]. In this type of stretching, a limb is moved near to its end point ROM and is typically maintained at this position for 15 to 60 s [5]. Transient increases in joint ROM following static stretching are due to a modified stretch sensation [6] and an increased compliance of the musculotendinous unit [7].

Loading characteristics of the stretching protocol, such as total stretch duration per muscle group, the duration of each stretching bout, rest between stretches, stretch intensity, and the muscle examined, influence acute joint ROM increases [8–11]. Previous research has reported that the total stretch duration is more important for joint ROM enhancement irrespective of whether stretching was performed in a continuous or intermittent manner [12,13]. In contrast, recent studies have found that different stretching modes (e.g., intermittent or continuous) induce dissimilar ROM changes [14,15]. Trajano, Nosaka, Seitz, and Blazevich, [14] compared an intermittent (5 × 1 min stretches with 15 s rest intervals between stretches) to a 5 min continuous stretching protocol, in young healthy adults and reported that only the continuous stretching protocol increased ankle angle during dorsiflexion, possibly due to a greater creep effect. In another study in elite male gymnasts, Bogdanis et al. [15] examined the effect of two different stretching protocols (3 × 30 s with 30 s rest, vs. 90 s) of the same total duration on hip and knee joint ROM. In that study it was found that intermittent and continuous stretching protocols induced similar increases in hip (+2.9° vs. +3.6°, $p = 0.001$, respectively) and knee joint ROM (+5.1° vs. +6.1°, $p = 0.001$, respectively) [15]. Such brief stretching durations (~30 s) are typically used in sports, since a typical warm-up routine includes 1–3 sets of shorter duration stretches (15–30 s) interspersed with rest intervals of equal duration, while the contra-lateral limb is being stretched [16,17]. Participants flexibility training experience may partly explain the discrepant results between studies. Along this line, Blazevich et al. [18] examined neuromuscular factors affecting maximum stretch limit and found that participants with a larger range of motion showed less resistance to stretch compared to less flexible participants due to a greater stretch tolerance. Cross-sectional studies comparing flexibility trained subjects (e.g., ballet dancers and rhythmic gymnasts) to controls or athletes from other sports also reported larger joint ROM in flexibility trained compared to untrained participants, as well as differences in muscle architecture [19,20].

Prolonged stretching durations (60–90 s or more) are used to maximize increases in ROM [4], facilitate recovery from injuries [21], prevent muscle mass loss in clinical conditions [22], and enhance performance in complex athletic tasks [23,24]. However, evidence is ambiguous regarding the effectiveness of performing static stretching in an intermittent or a continuous manner [14,15], especially when using stretching bouts lasting <60 s, as typically applied in sports. Also, the interaction of static stretching type and flexibility training background is largely unexplored. To this end, we aimed to examine changes in hip joint flexion angle after an intermittent (6 × 30 s with 30 s rest) or a continuous (180 s) static stretching protocol of equal total duration between athletes with different flexibility backgrounds. Hamstring muscles were chosen as they are important contributors to the work done at joints during explosive leg extensions [25], their extensibility determines hip and lumbar spine joint excursion [26], and a lack of hamstring flexibility is correlated to muscle injury [27]. It was hypothesized that flexibility trained athletes would respond differently to less-flexibility-trained counterparts, to an intermittent or continuous protocol of the same total duration.

2. Materials and Methods

2.1. Participants

Fourteen artistic and rhythmic gymnastics female college gymnasts were compared to thirteen female team sports athletes. All gymnasts trained 5–6 times per week (~180 min per training session) while team sports athletes competed in volleyball, basketball, and handball, and trained 2–3 times per week for 60–90 min. Gymnastics and team sports include a variety of locomotion activities using body weight, however gymnastics training additionally incorporates systematic daily flexibility training (~ 30–45 min), while team sports training includes less than 10 min of stretching exercises per training [28]. Participants anthropometric characteristics are shown in Table 1. Participants were healthy and did not report a lower limb injury for the past 6 months.

Table 1. Anthropometric characteristics of the participants (means ± standard deviation).

Characteristics	Artistic and Rhythmic Gymnasts (n = 14)	Team Sports Athletes (n = 13)	p
Age (y)	20.64 ± 2.68	19.15 ± 3.29	0.207
Training experience (y)	11.64 ± 2.34	7.23 ± 4.66	0.004
Height (m)	166.64 ± 5.29	168.77 ± 6.53	0.364
Body mass (kg)	55.21 ± 4.59	59.77 ± 4.90	0.020
Body Mass Index (kg/m^2)	19.86 ± 1.04	20.99 ± 1.46	0.031

2.2. Ethical Considerations

The study was approved by Institutional Ethics Committee (registration number: 1040, 14/02/2018). The design and conduct of the study was in accordance with the Helsinki declaration. Prior to the start of the study and after pertinent information of the procedures and potential risks involved were explained, participants signed a consent form.

2.3. Experimental Design

Participants performed one familiarization and one main testing session. During the familiarization session, participants' height and body mass were measured and they were familiarized with the testing procedures. The main testing session took place one week later and no intense exercise or stretching was allowed in the 48 h preceding testing. In order to stretch the hip extensors, the straight leg raise maneuver was performed to the point of discomfort, and it was applied on the one leg of the same individual in a continuous (180 s) and on the other leg in an intermittent manner (6 × 30 s with 30 s of rest in between).

In the main testing session, participants' hip flexion angle, was assessed using the straight leg raise test, in two conditions: (a) immediately after warm-up, and (b) following stretching intervention. The warm-up included 5 min of jogging at a moderate intensity (50%–60% of maximal heart rate). During testing, participants performed with the one leg the intermittent protocol and with the other leg the continuous stretching protocol. The assignment of the stretching type and the order of legs was done in a random and counterbalanced order. A schematic representation of the study protocol is shown in Figure 1.

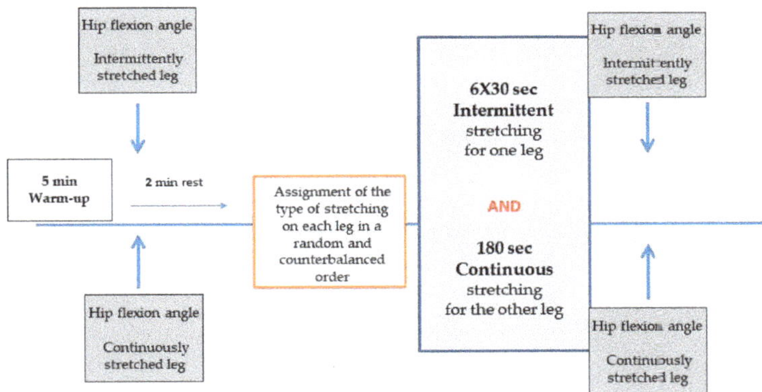

Figure 1. Schematic diagram of the study protocol.

2.4. Static Stretching Procedure and Range of Motion Measurements

The straight leg raise maneuver was chosen as a valid and reliable test of the extensibility of the hamstrings [29]. Furthermore, participants were familiar with this stretching movement because they used it in their training sessions. The static stretching protocols were applied and controlled by the same examiners. The straight leg raise was performed from a supine position on a physiotherapy bed, with the knee locked and the lower back flat on the bed (Figure 2). The lower back and the thigh of the non-stretched leg were stabilized with medical straps in order to prevent pelvic rotation. The participant's head was not supported by pillow. The examiner grasped the participant's heel of the tested leg with the one hand while with the other hand maintained the knee in an extended position. Slowly the examiner raised the stretched leg by flexing the hip. At the point of discomfort, the examiner maintained the stretch intensity for 30 s. The stretching movement was repeated for five more times, interspersed with 30 s of rest each time. During continuous stretching, the same procedure was followed and when the point of discomfort was reached, participants were instructed and verbally encouraged to maintain the stretch intensity for 3 min in order to induce the largest stretch they were willing to tolerate (Figure 2).

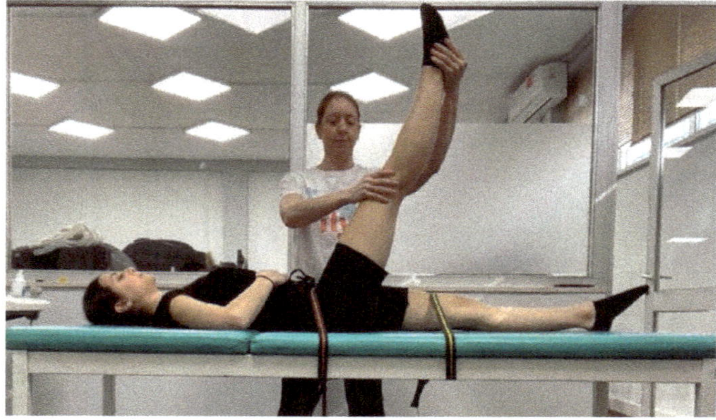

Figure 2. Straight leg raise movement.

Researchers supervised the position during testing, ensuring that both legs were straight, and the athletes kept a correct body alignment. Three anatomical markers were placed on hip (trochanterion), knee (femur–tibia joint line), and ankle (lateral malleolus) in order to analyze the images of hip flexion angle. A digital camera (Casio Exilim Pro EX-F1, Shibuya, Tokyo, Japan) placed perpendicular to the plane of motion of each leg, and aligned with the center of the hip joint, was used to record the position of the markers. Hip flexion angle was calculated using free software (Tracker 4.91 © 2016 Douglas Brown, Open Source Physics, Aptos, CA, USA). Straight leg raise angle was defined as the angle created by the intersection between horizontal and the line joining the hip, knee, and ankle markers. Intra-class correlation coefficients for hip angle was 0.96 (95% Confidence Intervals (CI): 0.81–0.99, $p < 0.001$). Participants gave feedback on stretch intensity to ensure that stretch achieved the point of discomfort (rating 80–90 to 100). Stretch intensity was indicated by the participants on a visual analogue scale used in previous studies, rated 0 ("no stretch discomfort at all") to 100 ("maximal stretch discomfort") [30].

2.5. Statistical Procedures

Descriptive statistics were calculated. Kolmogorov–Smirnov test checked for normality of data distribution. Unpaired *t*-test examined differences between groups in anthropometry. A three-way mixed model analysis of variance (ANOVA; time × stretching protocol × group) with time (rest vs. stretch) and stretching protocol (intermittent vs. continuous) as within-subjects factors and group (gymnasts vs. team sports athletes) as a between-subjects factor was conducted to examine the effect of stretching on hip flexion angle. Tukey's post-hoc test was performed when a significant main or interaction effect was observed ($p < 0.05$). Effect sizes for pairwise comparisons were calculated by Cohen's *d* [31]. Test–retest reliability was assessed by calculating the intra-class correlation coefficients (ICCs). Statistical significance was set at $p < 0.05$. Statistical analyzes were conducted using SPSS (SPSS Statistics Version 25.0, IBM corporation, Armonk, NY, USA).

3. Results

Hip Flexion Angle

There was a 3-way interaction of time x stretching protocol x group ($p = 0.041$, $\eta^2 = 0.157$). As shown in Figure 3, the intermittent stretching protocol significantly increased hip flexion angle in gymnasts and team sports athletes compared to baseline, by 6% and 13% (+8.2° ± 6.3°, 95% CI = +5.5 to +11.0°, $p = 0.029$, Cohen's $d = 0.43$ and +11.40° ± 16.3°, 95% CI = +2.6° to −20.2°, $p = 0.001$, Cohen's $d = 0.97$, respectively). In contrast, the continuous stretching protocol resulted in an increase in hip flexion only in gymnasts by 6%, while no increase was observed in team sport athletes (+7.7° ± 5.5° $p = 0.047$, 95% CI= +5.2° to +10.2°, Cohen's $d = 0.41$ vs, +1.0° ± 4.5°, $p = 0.99$ 95% CI = −1.1° to +3.1°, Cohen's $d = -0.09$, respectively; see Figure 3).

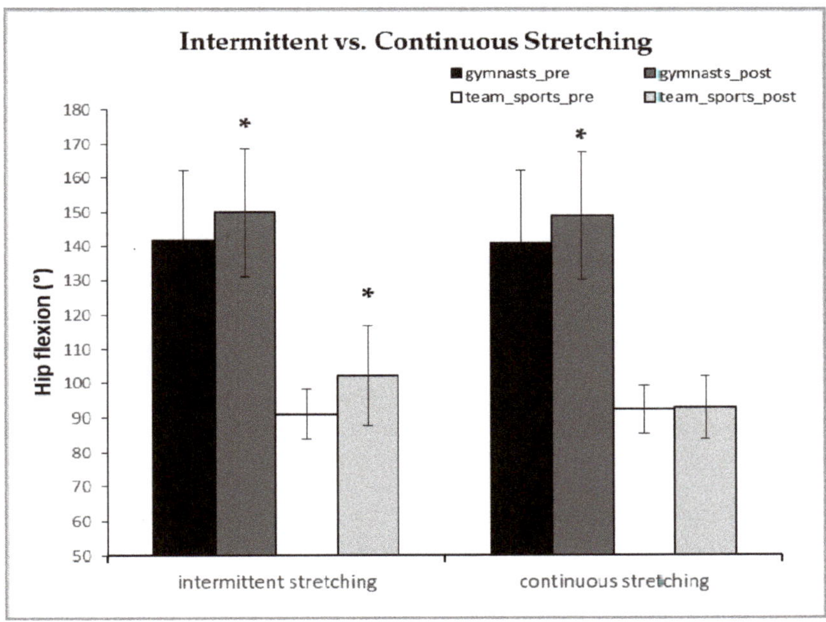

Figure 3. Hip flexion angle of gymnasts and team sport athletes during the intermittent and the continuous static stretching pre- and post-stretching intervention. Data are mean ± standard deviations of the mean. * $p < 0.01$ and $p < 0.05$ from pre-stretching.

4. Discussion

This study examined changes in hip joint flexion angle following an intermittent (6 × 30 s with 30 s rest) or a continuous (180 s) static stretching protocol of equal total duration between artistic and rhythmic gymnasts and team sports athletes. The main finding of this study was that both stretching protocols equally increased hip flexion angle in artistic and rhythmic female gymnasts while only intermittent stretching increased hip flexion angle in team sports athletes.

Stretching is commonly performed in sports and rehabilitation to enhance joint range of motion [5,9], increase the distance over which muscle force is applied [32], and prevent muscle injuries [33]. Despite a considerable number of studies examining the acute effect of stretching on different performance measures (i.e., muscle force and power) [9,19] the characteristics of the stretching protocols to induce optimal joint ROM increases are not sufficiently documented. Some previous studies have shown that intermittent or cyclic stretching is highly effective in increasing joint ROM. [34–36]. In line with these findings, the results of this study also indicate that intermittent stretching (6 × 30 s) increases hip flexion angle in gymnasts and team sport athletes by 6% and 13%, respectively, thus this type of stretching may be used as an effective method to enhance ROM in athletic populations. In a previous study, Cipriani et al. [13] compared the effect of a 6 × 10 s vs. 2 × 30 s stretches repeated twice daily on hip flexion and found significant increases in joint ROM irrespective of stretching duration. In an older study, Roberts et al. [37] examined the effect of 5 × 9 s vs. 3 × 15 s of static stretching on passive ROM of the lower extremities. The authors observed similar increases in passive joint ROM following both stretching protocols [37]. Collectively, studies comparing intermittent stretching protocols with different durations of each stretching bout, but equal total duration, found ROM enhancement regardless of the duration of each stretching. To this end, it was suggested that after 10 slow stretches, the passive tension of the muscle–tendon unit was reduced for any given length of the tissue, thus indicating tissue relaxation [38]. Konrad et al. [35] also reported decreased muscle stiffness immediately after, and up to 5 min following, a 5 min intermittent stretch (5 × 60 s) in healthy individuals. The efficacy of intermittent stretching may be due to the effect of rest between stretches, which allows recovery of the nervous, muscular, and metabolic systems, so that muscles can continue to extend against a pulling load [34].

In contrast, continuous stretching may not always be as effective as intermittent stretching although there are reports for the contrary [14]. However, in this study, continuous stretching did not have any effect on hip flexion ROM in team sports athletes, despite the fact that total duration was similar between the two protocols. Nordez et al. [36] examined the effect of constant (180 s) vs. cyclic stretching (6 × 30 s) on passive torque–angle curve of the hamstring muscles and reported that different mechanisms were operating depending upon the type of stretching performed. Although the present study did not examine neurophysiological and mechanical factors underpinning ROM changes, it is possible that maintaining a stretch position for 180 s may result in an increased reflexive activation in less trained in flexibility individuals even though the participants were instructed to relax during stretching. The exact mechanisms of the reflex muscle activity are not known, although input from spinal and supraspinal regions might be influential [18]. In addition, stretching transiently decreases muscle–blood flow in proportion to the applied tensile force of the stretch [39,40]. Thus, an ischemic response can occur during a passive muscle stretch due to an increase in intramuscular pressure, which in turn, may interfere with muscle activation [39]. Trajano et al. [14] found a two-fold higher muscle ischemia in a continuous (5 min) compared to an intermittent (5 × 1 min) stretching protocol. In contrast, rest intervals between stretching, allowed for blood reperfusion and minimized the decrease in HbO_2.

The fact that gymnasts' ROM increased equally with both stretching types may be due to their training background and the fact that prolonged static stretching is widely used in gymnastics. It is well established that flexibility training modifies stretch sensation (i.e., increases tolerance to the applied stretch) [41] so that individuals can tolerate higher levels of stretch for the same amount of perceived pain. It has also been reported that ballet dancers and rhythmic gymnasts have longer fascicles at rest and during stretching compared to controls [19,20], and this may be related to lower muscle tone, muscle stiffness, and resistance to stretching compared with team sports athletes. A previous study that examined changes in hip and knee joint ROM, following an intermittent (3 × 30 s with 30 s rest) or a continuous (90 s) static stretching protocol also reported that both stretching protocols similarly increased hip extension and knee flexion in international level male gymnasts [15]. Furthermore, flexibility trained individuals may also suffer less blood flow decrease during static stretching. For example, Otsuki et al. [40] examined muscle oxygenation and fascicle length during passive stretching between ballet dancers and controls and reported that ballet dancers demonstrated greater muscle extensibility without a concomitant reduction in muscle–blood volume and muscle oxygenation. The authors concluded that increased muscle extensibility attenuates muscle-tone-related vessels compression and indices of muscle ischemia during stretching [40]. Thus, a lower level of ischemia during continuous stretching in gymnasts compared to team athletes may partially explain their different responses. However, it is noteworthy that in relative terms, i.e., as a percentage of the resting value, team sport athletes increased their hip flexion ROM following intermittent stretching by 13% vs. 6% compared with gymnasts, albeit the absolute hip angles reached were far greater in gymnasts (Figure 3). This may be due to a possible "ceiling effect" of flexibility in gymnasts, whose ROM could have reached an absolute maximum and was possibly restricted by factors other than muscle extensibility (e.g., articular structures). The fact that gymnasts had longer training experience and were lighter compared to team sports athletes is a limitation that should be acknowledged. However, gymnastics training, as well as somatotype and anthropometric demands are unique and different from other sports.

5. Conclusions

In conclusion, both stretching types increased ROM in gymnasts, but only intermittent stretching was effective in team sports athletes. Thus, there is an interaction between stretching type and participants' flexibility training background that should be considered among other factors (e.g., training experience, somatotype) when examining the acute effects of static stretching on range of motion enhancement.

Author Contributions: Conceptualization, O.D., K.P., and G.C.B.; methodology, K.P. and I.P; investigation, K.P, V.G., and A.D.; data curation, K.P and I.P.; writing—original draft preparation, O.D., V.G., I.P., and A.D.; writing—review and editing, O.D. and G.C.B.; supervision, O.D. and G.C.B. All authors have read and agreed to the published version of the manuscript.

Funding: This research received no external funding.

Acknowledgments: The authors would like to thank the athletes for their dedicated time and collaboration.

Conflicts of Interest: The authors declare no conflict of interest.

References

1. Woods, K.; Bishop, P.; Jones, E. Warm-up and stretching in the prevention of muscular injury. *Sports Med.* **2007**, *37*, 1089–1099. [CrossRef]
2. Jamtvedt, G.; Herbert, R.D.; Flottorp, S.; Odgaard-Jensen, J.; Håvelsrud, K.; Barratt, A.; Oxman, A.D. A pragmatic randomized trial of stretching before and after physical activity to prevent injury and soreness. *Br. J. Sports Med.* **2010**, *44*, 1002–1009. [CrossRef] [PubMed]
3. Magnusson, P.; Renström, P. The European College of Sports Sciences position statement: The role of stretching exercises in sports. *Eur. J. Sport Sci.* **2006**, *6*, 87–91. [CrossRef]
4. Matsuo, S.; Suzuki, S.; Iwata, M.; Banno, Y.; Asai, Y.; Tsuchida, W.; Inoue, T. Acute effects of different stretching durations on passive torque, mobility, and isometric muscle force. *J. Strength Cond. Res.* **2013**, *27*, 3367–3376. [CrossRef] [PubMed]
5. Knudson, D. The biomechanics of stretching. *J. Exerc. Sci. Physiother.* **2006**, *2*, 3–12.
6. Magnusson, S.P. Passive properties of human skeletal muscle during stretch maneuvers. *Scand. J. Med. Sci. Sports* **1998**, *8*, 65–77. [CrossRef]
7. Morse, C.I.; Degens, H.; Seynnes, O.R.; Maganaris, C.N.; Jones, D.A. The acute effect of stretching on the passive stiffness of the human gastrocnemius muscle tendon unit. *J. Physiol.* **2008**, *586*, 97–106. [CrossRef]
8. Apostolopoulos, N.; Metsios, G.S.; Flouris, A.D.; Koutedakis, Y.; Wyon, M.A The relevance of stretch intensity and position- a systematic review. *Front. Psychol.* **2015**, *6*, 1128. [CrossRef]
9. Behm, D.G.; Blazevich, A.J.; Kay, A.D.; Mc Hugh, M. Acute effects of muscle stretching on physical performance, range of motion, and injury incidence in healthy active individuals: A systematic review. *Appl. Physiol. Nutr. Metab.* **2015**, *41*, 1–11. [CrossRef]
10. Freitas, S.R.; Andrade, R.J.; Larcoupaille, L.; Mill-Homens, P.; Nordez, A. Muscle and joint responses during and after static stretching performed at different intensities. *Eur. J. Appl. Physiol.* **2015**, *115*, 1263–1272. [CrossRef]

11. Lima, C.D.; Brown, L.E.; Wong, M.A.; Leyva, W.D.; Pinto, R.S.; Cadore, E.L.; Ruas, C.V. Acute Effects of Static vs. Ballistic Stretching on Muscular Fatigue Between Ballet Dancers and Resistance Trained Women. *J. Strength Cond. Res.* **2016**, *30*, 3220–3227. [CrossRef] [PubMed]
12. Chan, S.P.; Hong, Y.; Robinson, P.D. Flexibility and passive resistance of the hamstrings of young adults using two different static stretching protocols. *Scand. J. Med. Sci. Sports* **2001**, *11*, 81–86. [CrossRef] [PubMed]
13. Cipriani, D.; Abel, B.; Pirrwitz, D. A comparison of two stretching protocols on hip range of motion: Implications for total daily stretch duration. *J. Strength Cond. Res.* **2003**, *17*, 274–278. [CrossRef] [PubMed]
14. Trajano, G.S.; Nosaka, K.; Seitz, L.B.; Blazevich, A.J. Intermittent stretch reduces force and central drive more than continuous stretch. *Med. Sci. Sports Exerc.* **2014**, *46*, 902–910. [CrossRef] [PubMed]
15. Bogdanis, G.C.; Donti, O.; Tsolakis, C.; Smillios, I.; Bishop, D.J. Intermittent but not continuous static improves subsequent vertical jump performance in flexibility-trained athletes. *J. Strength Cond. Res.* **2019**, *33*, 203–210. [CrossRef] [PubMed]
16. Behm, D.G.; Chaouachi, A. A review of the acute effects of static and dynamic stretching on performance. *Eur. J. Appl. Physiol.* **2011**, *111*, 2633–2651. [CrossRef] [PubMed]
17. Simenz, C.J.; Dugan, C.A.; Ebben, W.P. Strength and conditioning practices of National Basketball Association strength and conditioning coaches. *J. Strength Cond. Res.* **2005**, *19*, 495–504. [CrossRef]
18. Blazevich, A.J.; Cannavan, D.; Waugh, C.M.; Fath, F.; Miller, S.C.; Kay, A.D. Neuromuscular factors influencing the maximum stretch limit of the human plantar flexors. *J. Appl. Physiol.* **2012**, *113*, 1446–1455. [CrossRef]
19. Donti, O.; Panidis, I.; Terzis, G.; Bogdanis, G.C. Gastrocnemius Medialis Architectural Properties at Rest and During Stretching in Female Athletes with Different Flexibility Training Background. *Sports* **2019**, *7*, 39. [CrossRef]
20. Moltubakk, M.M.; Magulas, M.M.; Villars, F.O.; Seynnes, O.R.; Bojsen-Møller, J. Specialized properties of the triceps surae muscle-tendon unit in professional ballet dancers. *Scand. J. Med. Sci. Sports* **2018**, *28*, 2023–2034. [CrossRef]
21. Esposito, F.; Limonta, E.; Cè, E. Passive stretching effects on electromechanical delay and time course of recovery in human skeletal muscle: New insights from an electromyographic and mechanomyographic combined approach. *Eur. J. Appl. Physiol.* **2011**, *111*, 485–495. [CrossRef] [PubMed]
22. Lowe, D.A.; Alway, S.E. Animal models for inducing muscle hypertrophy: Are they relevant for clinical applications in humans? *J. Orthop. Sports Phys. Ther.* **2002**, *32*, 36–43. [CrossRef] [PubMed]
23. Donti, O.; Tsolakis, C.; Bogdanis, G.C. Effects of baseline levels of flexibility and vertical jump ability on performance following different volumes of static stretching and potentiating exercises in elite gymnasts. *J. Sports Sci. Med.* **2014**, *13*, 105–113. [PubMed]
24. Moltubakk, M.M.; Eriksrud, O.; Paulsen, G.; Seynnes, O.R.; Bojsen-Møller, J. Hamstrings functional properties in athletes with high musculo-skeletal flexibility. *Scand. J. Med. Sci. Sports* **2016**, *26*, 659–665. [CrossRef] [PubMed]
25. Jacobs, R.; Bobbert, M.F.; Van Ingen, S.G.J. Mechanical output from individual muscles during explosive leg extensions: The role of biarticular muscles. *J. Biomech.* **1996**, *29*, 513–523. [CrossRef]
26. Johnson, E.N.; Thomas, J.S. Effect of hamstring flexibility on hip and lumbar spine joint excursions during forward-reaching tasks in participants with and without low back pain. *Arch. Phys. Med. Rehabil.* **2010**, *91*, 1140–1142. [CrossRef]
27. Worrell, T.W.; Perrin, D.H. Hamstring muscle injury: The influence of strength, flexibility, warm-up, and fatigue. *J. Orthop. Sports Phys. Ther.* **1992**, *16*, 12–18. [CrossRef]
28. Donti, O.; Bogdanis, G.C.; Kritikou, M.; Donti, A.; Theodorakou, K. The relative contribution of physical fitness to the technical execution score in youth rhythmic gymnastics. *J. Hum. Kinet.* **2016**, *51*, 143–152. [CrossRef]
29. Heyward, V.H. *Advanced Fitness Assessment and Exercise Prescription*, 4th ed.; Human Kinetics: Champaign, IL, USA, 2005; pp. 230–240.
30. Behm, D.G.; Kibele, A. Effects of differing intensities of static stretching on jump performance. *Eur. J. Appl. Physiol.* **2007**, *101*, 587–594. [CrossRef]
31. Cohen, J. A power primer. *Psychol. Bull.* **1992**, *112*, 155–159. [CrossRef]
32. Falsone, S. *High Performance Training for Sports*; Human Kinetics: Champain, IL, USA, 2014; pp. 61–70.
33. Witvrouw, E.; Mahieu, N.; Danneels, L.; McNair, P. Stretching and injury prevention. *Sports Med.* **2004**, *34*, 443–449. [CrossRef] [PubMed]

34. Magnusson, S.P.; Aagard, P.; Simonsen, E.; Bojsen-Møller, F. A biomechanical evaluation of cyclic and static stretch in human skeletal muscle. *Int. J. Sports Med.* **1998**, *19*, 310–316. [CrossRef] [PubMed]
35. Konrad, A.; Reiner, M.M.; Thaller, S.; Tilp, M. The time course of muscle-tendon properties and function responses of a five-minute static stretching exercise. *Eur. J. Sport Sci.* **2019**, *19*, 1195–1203. [CrossRef] [PubMed]
36. Nordez, A.; McNair, P.J.; Casari, P.; Cornu, C. The effect of angular velocity and cycle on the dissipative properties of the knee during passive cyclic stretching: A matter of viscosity or solid friction. *Clin. Biomech.* **2009**, *24*, 77–81. [CrossRef]
37. Roberts, J.M.; Wilson, K. Effect of stretching duration on active and passive range of motion in the lower extremity. *Br. J. Sports Med.* **1999**, *33*, 259–263. [CrossRef]
38. Taylor, D.C.; Brooks, D.E.; Ryan, J.B. Viscoelastic characteristics of muscle: Passive stretching versus muscular contractions. *Med. Sci. Sports Exerc.* **1997**, *29*, 1619–1624. [CrossRef]
39. Özkaya, N.; Nordin, M. Equilibrium, motion, and deformation. In *Fundamentals of Biomechanics*, 2nd ed.; Springer: Berlin, Germany, 1999; pp. 221–236. [CrossRef]
40. Otsuki, A.; Fujita, E.; Ikegawa, S.; Kuno-Mizumura, M. Muscle oxygenation and fascicle length during passive muscle stretching in ballet-trained subjects. *Int. J. Sports Med.* **2011**, *32*, 496–502. [CrossRef]
41. Weppler, C.H.; Magnusson, S.P. Increasing muscle extensibility: A matter of increasing length or modifying sensation? *Phys. Ther.* **2010**, *90*, 438–449. [CrossRef]

Article

The Effects of Postprandial Resistance Exercise on Blood Glucose and Lipids in Prediabetic, Beta-Thalassemia Major Patients

Kalliopi Georgakouli [1,2], Alexandra Stamperna [1], Chariklia K. Deli [1], Niki Syrou [1], Dimitrios Draganidis [1], Ioannis G. Fatouros [1] and Athanasios Z. Jamurtas [1,*]

1 Department of Physical Education and Sport Science, University of Thessaly, Karies, 42100 Trikala, Greece; kgeorgakouli@gmail.com (K.G.); alexstaberna@gmail.com (A.S.); delixar@pe.uth.gr (C.K.D.); nikisyrou@uth.gr (N.S.); dimidraganidis@gmail.com (D.D.); ifatouros@uth.gr (I.G.F.)

2 Department of Nutrition & Dietetics, University of Thessaly, Karies, 42100 Trikala, Greece

* Correspondence: ajamurt@pe.uth.gr; Tel.: +30-243-104-7054

Received: 29 January 2020; Accepted: 21 April 2020; Published: 26 April 2020

Abstract: Insulin resistance and diabetes mellitus are common consequences of iron overload in the pancreas of beta-thalassemia major (BTM) patients. Moreover, postprandial blood glucose elevations are linked to major vascular complications. The purpose of this study was to investigate the effects of a bout of acute resistance exercise following breakfast consumption of glucose and fat on the metabolism in prediabetic, BTM patients. Six patients underwent two trials (exercise and control) following breakfast consumption (consisting of approximately 50% carbohydrates, 15% proteins, 35% fat), in a counterbalanced order, separated by at least three days. In an exercise trial, patients performed chest and leg presses (3 sets of 10 repetitions maximum/exercise), while in the control trial they rested. Blood samples were obtained in both trials at: pre-meal, 45 min post-meal (pre-exercise/control), post-exercise/control, 1 h post-exercise/control, 2 h post-exercise/control and 24 h post-exercise/control. Blood was analysed for glucose and lipids (total cholesterol, High Density Lipoprotein-cholesterol, Low Density Lipoprotein-cholesterol, triglycerides). Blood glucose levels increased significantly 45 min following breakfast consumption. Blood glucose and lipids did not differ between trials at the same time points. It seems that a single bout of resistance training is not sufficient to improve blood glucose and fat levels for the subsequent 24-h post-exercise period in prediabetic, BTM patients.

Keywords: haemoglobin; diabetes; fitness; cardiovascular health; nutrition

1. Introduction

Prediabetes is as a state of intermediate hyperglycaemia where a person has impaired fasting glucose, impaired glucose tolerance or a combination of the two [1]. Although glucose levels in prediabetes are not yet high enough for a diagnosis of Diabetes Mellitus (DM), prediabetes is associated with macrovascular and microvascular complications of DM, including nephropathy, small fibre neuropathy, retinopathy and coronary artery disease. Moreover, individuals with prediabetes are at high risk of developing DM [2].

Lifestyle interventions have been shown to reduce the risk of DM in adults with prediabetes, and they should be an essential part of the management of this condition. The main problem in glycaemic control is the peak of glucose 1–2 h after a meal, i.e., postprandial hyperglycaemia. Exercise increases contraction-mediated glucose uptake resulting in reduced postprandial hyperglycaemia and has been proposed as an effective way to improve glucose control in individuals with type 2 DM. Indeed, the timing of exercise relative to meal consumption may play a role in glycaemic control. The limited available data indicate that postprandial (consumption following dinner) exercise may be more beneficial than preprandial exercise in type 2 DM patients, and both aerobic and resistance training have been shown to be effective [3]. In addition, postprandial resistance exercise improves triglyceride levels, another risk factor for cardiovascular disease in type 2 DM [4]. Thus, postprandial resistance exercise may be an effective means of better glycaemic control and a lower risk of cardiovascular disease in individuals with an abnormal glucose metabolism. The optimal postprandial exercise timing and prescription are yet to be defined.

Beta-thalassemia major (BTM) is an inherited haemoglobin disorder that manifests within the first few months of life with ineffective erythropoiesis and chronic haemolytic anaemia, and frequent blood transfusions are required. There is no physiological mechanism to remove the excess iron load resulting from regular blood transfusions, while ineffective erythropoiesis increases intestinal iron absorption. Both processes induce iron accumulation in reticuloendothelial cells and parenchymal tissues that can cause progressive damage in multiple organs [5]. Iron accumulation in the pancreatic islets induces insulin resistance and reduced early insulin secretion, often resulting in DM in BTM patients [6]. Although the pathophysiological mechanisms leading to the development of DM are still unclear, it is most likely linked to the reversible oxidation and reduction of iron. This property renders iron potentially hazardous due to its ability to participate in the generation of reactive oxygen species [7], while pancreatic islets are susceptible to oxidative damage as they almost exclusively rely on the mitochondrial metabolism of glucose for glucose-induced insulin secretion and they also have a low antioxidant defence system [8]. Exercise is thought to provide various beneficial health effects in various metabolic disorders; however, research on the effects of exercise in BTM patients is non-existent. This could be due to the fact that BTM patients often manifest exercise intolerance and fatigue mediated by anaemia and iron-mediated cardiotoxicity [9].

This study was therefore designed to investigate whether postprandial resistance exercise can influence changes in glucose and lipid metabolism in prediabetic, BTM patients. Based on previous literature on the effects of postprandial exercise on health parameters in prediabetics, we hypothesized that an acute bout of resistance exercise 45 min following breakfast consumption would attenuate the blood glucose response and improve lipid profiles throughout the subsequent 24-h post-exercise period.

2. Materials and Methods

2.1. Participants

Six prediabetic, BTM patients (3 men, 3 women; age: 39.5 ± 4.6 years) volunteered to participate in this study. All of them had their doctor's permission to participate, were informed about the study protocol, filled a medical history questionnaire and signed an informed consent form. Due to transfusions, these patients had suppressed their autologous hematopoiesis. Haemoglobin in BTM patients comes mainly from frequent transfusions and therefore has a normal O2 affinity. Procedures were in accordance with the 1975 Declaration of Helsinki (2000) and approval was obtained from the Institutional Review Board of the Department of Physical Education and Sport Science, University of Thessaly (protocol number 1076). Moreover, the study was registered at ClinicalTrials.gov as NCT03889977 [10].

Inclusion criteria:

- BTM patients requiring regular blood transfusions.
- Confirmed prediabetes with patients fulfilling one if the three following criteria: (a) impaired fasting glucose (IFG) (fasting plasma glucose (FPG) of 6.1–6.9 mmol/L), or (b) impaired glucose tolerance (IGT) (plasma glucose of 7.8–11.0 mmol/L, 2 h following ingestion of 75 g of oral glucose load) or (c) a combination of the two, based on a 2 h oral glucose tolerance test [1].
- Age group of 25 to 55.

Exclusion criteria:

- Hypertension.

- Injuries.
- Any other serious complications of BTM that could contraindicate participation to exercise.

Moreover, volunteers were instructed to avoid lifestyle changes and strenuous or unusual physical activity for at least two days before each visit to the laboratory.

2.2. Experimental Design

All participants reported to the laboratory in the morning (9:00–10:00 a.m.) for physiological measurements (blood pressure, resting heart rate, body composition analysis through DEXA test) and determination of one repetition maximum (1RM). Moreover, they were asked to record their diet for two days before their next visit and followed the same diet before the third and last visit.

In a randomized within-subject design, participants underwent two trials (exercise, ExT and control, CoT) 45 min following breakfast consumption (consisting of approximately 50% carbohydrates, 15% proteins, 35% fat), in a counterbalanced order, separated by at least three days.

In each one of the experimental trials, participants reported to the laboratory in the morning (9:00–10:00 a.m.) after overnight fasting (~10 h) and a blood sample was drawn. Then they were provided with a standard breakfast that they had to consume in 10 min. Forty-five min later, another blood sample was drawn. In the ExT, the participants initially performed a warm-up for 5–10 min consisting of 8–10 repetitions using a light weight. Then they performed chest and leg press exercises (3 sets of 10 repetitions at 70% of their 1RM, each), and they finished the workout by performing stretching exercises for approximately 5 min of the two major muscle groups used during the workout. In the CoT, participants rested for the same duration. Blood samples were obtained immediately after ExT and CoT, as well as 1 h, 2 h and 24 h following each trial. Water consumption throughout trials was ad libitum.

2.3. Anthropometric and Physiological Characteristics

Body height was measured with a precision of 0.1 cm and body weight with a precision of 0.1 kg (Beam Balance, Seca, Birmingham, UK), with the participants lightly dressed and barefoot. Body fat percentage was estimated by dual-energy X-ray absorptiometry (DEXA) (Lunar DPX NT, GE Healthcare, Chalfont St. Giles, Buckinghamshire, UK). Blood pressure (BP) was measured with a manual sphygmomanometer (FC-101 Aneroid Sphygmomanometer; Focal Corporation, Kashiwa, Japan) after 5 min of seated rest.

2.4. Blood Collection and Handling

In total, blood samples were obtained from a forearm vein with participants in a seated position at the following time-points: pre-meal, 45 min post-meal (pre-ExT/CoT), immediately post- ExT/CoT, 1 h post-ExT/CoT, 2 h post-ExT/CoT, 24 h post- ExT/CoT. Blood samples were obtained after 10 min rest, except for immediately post-ExT, where blood was obtained immediately following exercise. Blood was immediately transferred to a tube containing Ethylene diamine tetraacetic acid (EDTA) and centrifuged in order to obtain the plasma which was stored in aliquots at −80 °C until the day of analysis. Plasma preparation has been described elsewhere [11].

2.5. Blood Analysis

Samples underwent only one freeze-thaw cycle and each parameter was measured in duplicates. Plasma glucose and lipids (total cholesterol, HDL-cholesterol, LDL-cholesterol, triglycerides) were determined using a biochemical analyser (Clinical Chemistry Analyzer Z 1145; Zafiropoulos Diagnostica S.A., Koropi, Greece). LDL-cholesterol was calculated using the Friedewald equation [12].

2.6. Statistical Analysis

Preliminary power analysis performed with G*Power [13] showed that the minimum required sample size was 6 (with a probability error of 0.05, a statistical power of 80% and an effect size of 0.5).

Normality was checked using the Shapiro–Wilk test. Since some variables did not follow normal distribution, nonparametric statistics were used for the analyses. The Friedman analysis of variance by ranks test was performed to determine the time effects, accompanied by the Wilcoxon signed-rank test to perform pairwise comparisons. Differences between trials were examined using the Mann–Whitney U test (U values). Differences between genders for areas under the curve (AUC) and triglycerides was assessed by a 2 × 2 (gender by condition) ANOVA. The level of statistical significance was set at $p < 0.05$. Data are presented as means ± SD. Statistical analysis was conducted with IBM SPSS Version 19.0 (IBM Corp., Armonk, NY, USA).

Moreover, the incremental areas under the curves (AUCs) were measured for plasma glucose and triglycerides with GraphPad Prism version 5.0 (GraphPad Software, San Diego, CA, USA). Differences in AUCs between trials were examined using the Mann–Whitney U test (U values).

3. Results

3.1. Anthropometric and Physiological Characteristics

Anthropometric and physiological characteristics of the participants did not differ between the ExT and CoT (Table 1). The mean body mass index (BMI) indicated that men were overweight and women were normal weight. According to the International Diabetes Federation, the ethnic group specific cut-point for waist circumference (WC) is 94 cm and 80 cm for European men and women, respectively [14]. WC of the participants in the present study exceeded this cut-point, indicating central obesity.

Table 1. Anthropometric and physiological characteristics before exercise (ExT) and control trial (CoT).

Variable	ExT	CoT
BM (kg)	66.0 ± 16.6	65.5 ± 16.2
BMI (kg/m^2)	24.2 ± 5.4 (men: 25.7 ± 7.2; women: 22.6 ± 3.8)	24.0 ± 5.2 (men: 25.6 ± 6.8; women: 22.4 ± 3.7)
%BF	37.6 ± 5.1 (men: 25.7; women: 22.6)	37.6 ± 5.1 (men: 25.7; women: 22.6)
WC (cm)	93.5 ± 12.2 (men: 101.7 ± 9.1; women: 85.3 ± 9.5)	93.0 ± 11.6 (men: 101.0 ± 7.9; women: 85.0 ± 9.0)
HC (cm)	97.5 ± 8.2 (men: 100.3 ± 8.5; women: 94.7 ± 8.5)	97.5 ± 8.2 (men: 100.3 ± 8.5; women: 94.7 ± 8.5)
WHR	0.96 ± 0.07 (men: 1.01 ± 0.03; women: 0.90 ± 0.02)	0.95 ± 0.06 (men: 1.00 ± 0.03; women: 0.90 ± 0.01)
RHR	79.0 ± 8.6	77.5 ± 9.2
SBP (mmHg)	104.5 ± 9.7	103.7 ± 10.3
DBP (mmHg)	67.5 ± 7.6	67.5 ± 7.6

BM: Body Mass; BMI: Body Mass Index; %BF: Body Fat percentage; WC: Waist Circumference; HC: Hip Circumference; WHR: Waist to Hip Ratio; RHR: Resting Heart Rate; SBP: Systolic Blood Pressure; DPB: Diastolic Blood Pressure.

3.2. Metabolic Parameters

In ExT, plasma glucose levels increased 45 min following breakfast consumption (%z = −2 20; $p = 0.028$), immediately following exercise (%z = −1.99; $p = 0.046$) and 1 h following exercise (%z = −1.99; $p = 0.046$) compared to baseline levels (before breakfast consumption). In CoT, plasma glucose levels were increased 45 min following breakfast consumption (%z = −2.20; $p = 0.028$) and immediately following rest (%z = −2.02; $p = 0.043$) compared to baseline levels (before breakfast consumption). Moreover, pairwise comparisons showed that there was no significant difference in glucose levels at any time point between trials (Figure 1). Table 2 indicates time related changes with significant increases observed at 45 min post-meal following the exercise and control trials, immediately post-exercise/control and 1 h post-exercise. Moreover, no difference in glucose area under the curve (AUC) between trials or gender was observed (Figure 2).

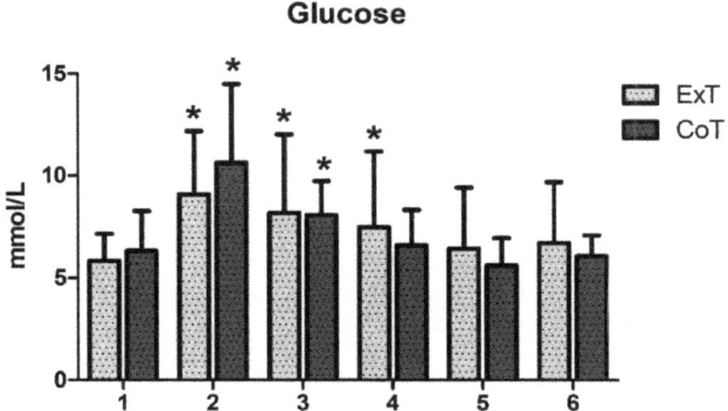

Figure 1. Changes in plasma glucose levels (mmol/L) following exercise (ExT) and control trial (CoT). Time points: (1) pre-meal; (2) 45 min post-meal (pre-exercise/control); (3) immediately post-exercise/control; (4) 1 hr post-exercise/control; (5) 2 h post-exercise/control; (6) 24 h post-exercise/control. *Significant difference from (1) at the same trial.

Table 2. Changes in plasma glucose (mmol/L) levels (mean ± SD) following exercise (ExT) and control trial (CoT). Time points: (1) pre-meal; (2) 45 min post-meal (pre-exercise/control); (3) immediately post-exercise/control; (4) 1 h post-exercise/control; (5) 2 h post-exercise/control; (6) 24 h post-exercise/control. * Significant difference from (1) at the same trial.

Time point	ExT	CoT
1	5.83 ± 1.33	6.32 ± 1.95
2	9.07 ± 3.11 *	10.62 ± 3.87 *
3	8.17 ± 3.85 *	8.07 ± 1.66 *
4	7.47 ± 3.71 *	6.60 ± 1.73
5	6.43 ± 2.99	5.61 ± 1.35
6	6.71 ± 2.99	6.07 ± 1.01

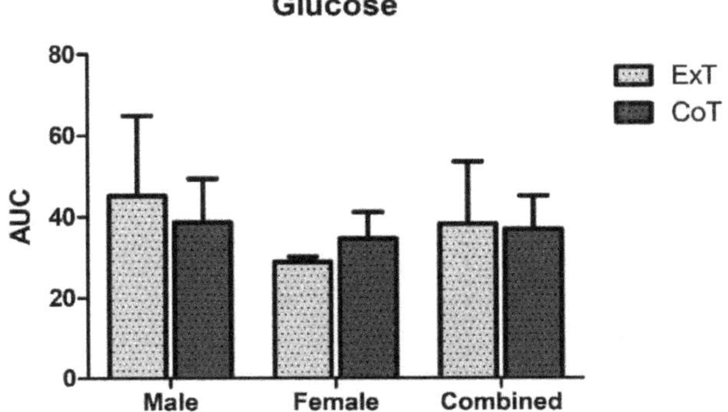

Figure 2. Glucose area under the curve (AUC) in mmol/L × 24h in exercise (ExT) and control trial (CoT).

Triglycerides, total cholesterol, HDL and LDL levels did not change at any time point and were similar in both conditions (Figure 3a–d). Moreover, no difference in triglycerides AUC between trials or gender was observed (Figure 4).

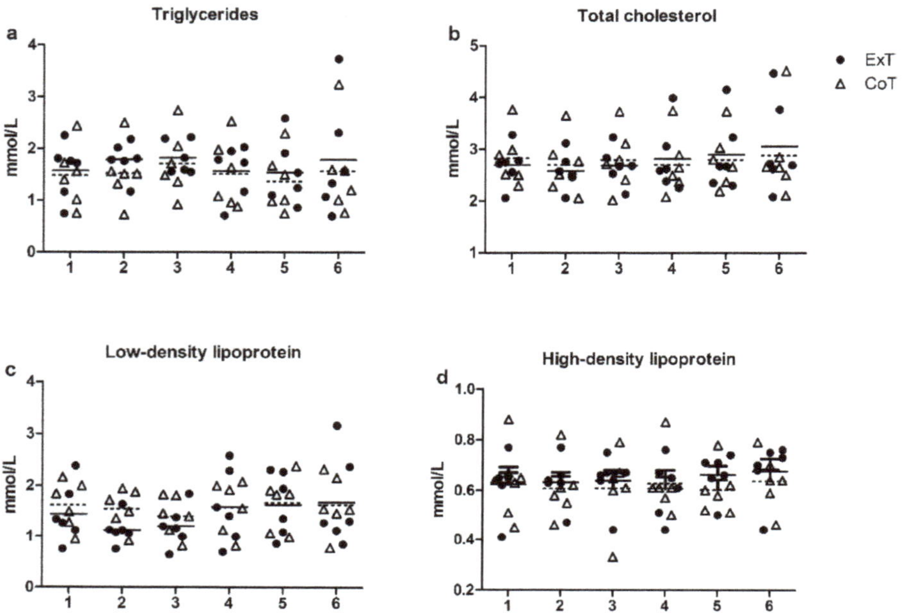

Figure 3. Changes in plasma lipid levels following exercise (ExT) and control trial (CoT): (**a**) Triglycerides; (**b**) Total cholesterol; (**c**) LDL: Low Density Lipoprotein; (**d**) HDL: High Density Lipoprotein. Time points: (1) pre-meal; (2) 45 min post-meal (pre-exercise/control); (3) immediately post-exercise/control; (4) 1 h post-exercise/control; (5) 2 h post-exercise/control; (6) 24 h post-exercise/control. Dotted lines represent the mean values of the Control trial.

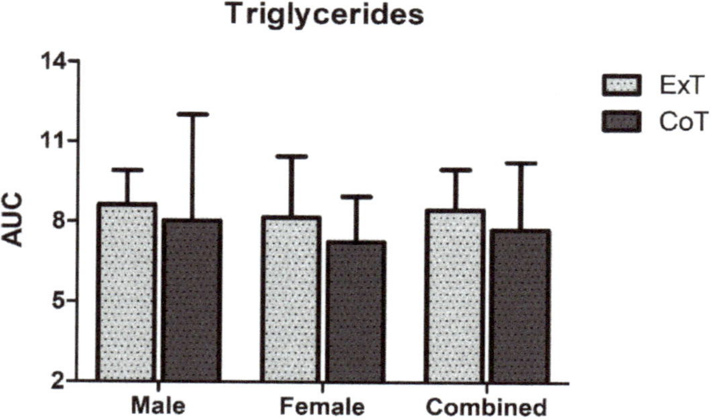

Figure 4. Triglycerides area under the curve (AUC) in mmol/L × 24h in exercise (ExT) and control trial (CoT).

4. Discussion

The main aim of this study was to examine whether postprandial resistance exercise can influence changes in blood glucose in prediabetic, BTM patients. Supplementary analyses of blood lipids were performed as these parameters are associated with glucose metabolism, DM and cardiovascular disease. To the authors' knowledge, this is the first study to test this hypothesis in this clinical population. Our findings suggest that an acute bout of postprandial (post-breakfast) resistance exercise is not a sufficient stimulus to (i) attenuate the blood glucose response, and (ii) change the lipids profile, throughout the subsequent 24-h post-exercise period.

Although research on the effects of resistance training on glycaemic control in type 2 DM patients is scarce, there is some data suggesting that this modality could be beneficial [15]. Moreover, the positive effects of a single session of aerobic exercise (50% maximum workload capacity) and a single session of resistance exercise (75% 1RM) on the 24-h average blood glucose levels and the 24-h prevalence of hyperglycaemia were found to be similar [16]. In the present study, we did not observe any change in blood glucose levels throughout the subsequent 24-h post-exercise period. It is possible that prediabetic, BTM patients do not have similar responses to those of type 2 DM patients as the underlining pathophysiology could lead to different results. Moreover, the high level of SD observed could be explained by the small sample. The population examined in our study has very unique characteristics and many factors (i.e., medication, frequency of blood transfusion, comorbidities) can lead to different responses. Moreover, SD was higher in Ext at time points 2–6, indicating that these unique characteristics may have an important role in blood glucose responses to exercise for up to 24-h. Besides, it has been reported that BTM patients often manifest exercise intolerance and fatigue mediated by anaemia and iron-mediated cardiotoxicity [9]. Thus more research comparing the effects of resistance exercise in prediabetic, BTM patients and DM patients is warranted.

The timing of exercise relative to meal consumption may also play a role in glycaemic control. The limited available data indicate that postprandial exercise may be more beneficial than preprandial exercise in type 2 DM patients [3]. Postprandial resistance exercise causes a greater reduction in glucose incremental area under the curve (iAUC) (reduction by 30%) compared to preprandial resistance exercise (reduction by 18%) [4]. In addition, postprandial resistance exercise improves triglyceride levels, another risk factor for cardiovascular disease in type 2 DM [4]. Thus, postprandial resistance exercise may be an effective means of better glycaemic control and lower risk of cardiovascular disease in individuals with abnormal glucose metabolism. Based on this, Borror et al. [3] proposed that resistance exercise should be performed following the largest meal of the day, 2 to 3 non-consecutive days per week, at intensities varying between 50% and 80% of 1RM, working the major muscle groups (1–4 sets of 8–15 repetitions/exercise) [17,18]. In the present study, we used an acute resistance exercise protocol at an intensity of 70% of 1RM and we found no change in blood glucose levels, glucose AUC, lipid levels or triglycerides AUC throughout the 24-h post-exercise period. The limited available data suggest that resistance exercise could have positive effects that are associated with increases in lean muscle mass and type II fibre type recruitment [15]. However, all aforementioned studies refer to acute postprandial exercise studies and it needs to be stated that it is of great importance to perform exercise training interventions to assess the long-term effects of exercise on not only the acute hyperglycaemia, but the long-term glycaemia control as well. Previous reports indicate that glycaemia in type 2 DM males and females is different with males exhibiting higher impaired fasting glycaemia and women impaired glucose tolerance [19,20]. The results from this study do not coincide with the aforementioned reports since there were no differences between males and females neither at rest nor following the glycaemic load (AUC) at rest and after exercise. However, it needs to be stated here that the sample size for males (n = 3) and females (n = 3) is rather small, this constitutes a limitation of the study and concrete conclusions should be made with caution.

Postprandial hypertriglyceridemia has also been linked to increased risk of cardiovascular disease [21]. In this study, blood lipids (triglycerides, total cholesterol, HDL and LDL) did not change at any time point and were similar in both trials. This was also evident in the lack of difference in triglycerides AUC between trials. This may be explained by the fact that the breakfast provided was low in saturated fat. A meal high in saturated fat increases blood triglyceride levels as well as indices of oxidative stress and inflammation, resulting in a worsening of endothelial dysfunction, vasoconstriction and systolic blood pressure [22,23]. Therefore, exercise following lunch or dinner could be more beneficial in terms of cardiovascular health.

Patients with abnormal glucose levels are often diagnosed with Metabolic Syndrome (MS), a multiple set of risk factors that confer an additional cardiovascular risk [24]. In this study, it was shown that men were overweight and women were of normal weight according to BMI, and that both genders had central obesity according to the WC cut-point [14]. The participants were also prediabetic and had low HDL levels, meeting the criteria for metabolic syndrome set by the International Diabetes Federation. As already mentioned, BTM causes various complications due to anaemia and iron overload, and the presence of MS is an additional cardiovascular risk, rendering BTM patients a clinical population with unique characteristics. Therefore, BTM patients may not respond to exercise in a typical manner and long-term interventions are required.

In conclusion, postprandial resistance exercise, especially following the largest meal of the day, could be an effective means of glycaemic control and could lower the risk of cardiovascular disease in DM patients [3]. The results of the present study do not indicate that a bout of acute resistance exercise 45 min following breakfast is a sufficient stimulus to improve blood glucose lipid levels throughout the subsequent 24-h post-exercise period. Future studies comparing the acute and chronic effects of resistance exercise in prediabetic, BTM patients and DM patients are needed.

Author Contributions: Conceptualization, A.Z.J.; methodology, A.Z.J., K.G. and C.K.D.; validation, A.Z.J.; formal analysis, K.G.; investigation, K.G., A.S., C.K.D. and N.S.; resources, A.Z.J. and A.S.; data curation, K.G. and A.Z.J.; writing—original draft preparation, K.G.; writing—review and editing, A.S., I.G.F. and A.Z.J.; visualization, K.G., D.D. and A.S.; supervision, A.Z.J. and I.G.F.; project administration, A.Z.J. and A.S.; funding acquisition, A.Z.J. and I.G.F. All authors have read and agreed to the published version of the manuscript.

Funding: This research received partial funding from the Postgraduate Program of Study "Exercise and Health: Testing & Prescription", Department of P.E. & Sport Science, University of Thessaly, Greece.

Conflicts of Interest: The authors declare no conflict of interest.

References

1. World Health Organization. *Definition and Diagnosis of Diabetes Mellitus and Intermediate Hyperglycemia: Report of a WHO/IDF Consultation*; World Health Organization: Geneva, Switzerland, 2006; pp. 1–50.
2. Bansal, N. Prediabetes diagnosis and treatment: A review. *World J. Diabetes* **2015**, *6*, 296–303. [CrossRef] [PubMed]
3. Borror, A.; Zieff, G.; Battaglini, C.; Stoner, L. The Effects of Postprandial Exercise on Glucose Control in Individuals with Type 2 Diabetes: A Systematic Review. *Sports Med.* **2018**, *48*, 1479–1491. [CrossRef] [PubMed]
4. Heden, T.D.; Winn, N.C.; Mari, A.; Booth, F.W.; Rector, R.S.; Thyfault, J.P.; Kanaley, J.A. Postdinner resistance exercise improves postprandial risk factors more effectively than predinner resistance exercise in patients with type 2 diabetes. *J. Appl. Physiol.* **2015**, *118*, 624–634. [CrossRef] [PubMed]
5. Taher, A.T.; Saliba, A.N. Iron overload in thalassemia: Different organs at different rates. Hematology. *Hematol. Am. Soc. Hematol. Educ. Program* **2017**, *2017*, 265–271. [CrossRef]
6. Cario, H.; Holl, R.W.; Debatin, K.M.; Kohne, E. Insulin sensitivity and beta-cell secretion in thalassaemia major with secondary haemochromatosis: Assessment by oral glucose tolerance test. *Eur. J. Pediatr.* **2003**, *162*, 139–146. [CrossRef]
7. Halliwell, B.; Gutteridge, J.M.C. Role of free radicals and catalytic metal ions in human disease: An overview. *Methods Enzymol.* **1990**, *186*, 1–85.
8. Tiedge, M.; Lortz, S.; Drinkgern, J.; Lenzen, S. Relation between antioxidant enzyme gene expression and antioxidative defense status of insulin-producing cells. *Diabetes* **1997**, *46*, 1733–1742. [CrossRef]
9. Sohn, E.Y.; Kato, R.; Noetzli, L.J.; Gera, A.; Coates, T.; Harmatz, P.; Keens, T.G.; Wood, J.C. Exercise performance in thalassemia major: Correlation with cardiac iron burden. *Am. J. Hematol.* **2013**, *88*, 193–197. [CrossRef]

10. ClinicalTrials.gov. US National Institutes of Health. Available online: http://www.clinicaltrials.gov (accessed on 14 January 2020).
11. Georgakouli, K.; Manthou, E.; Fatouros, I.G.; Deli, C.K.; Spandidos, D.A.; Tsatsakis, A.M.; Kouretas, D.; Koutedakis, Y.; Theodorakis, Y.; Jamurtas, A.Z. Effects of acute exercise on liver function and blood redox status in heavy drinkers. *Exp. Ther. Med.* **2015**, *10*, 2015–2022. [CrossRef]
12. Friedewald, W.T.; Levy, R.I.; Fredrickson, D.S. Estimation of the concentration of low-density lipoprotein cholesterol in plasma, without use of the preparative ultracentrifuge. *Clin. Chem.* **1972**, *18*, 499–502. [CrossRef]
13. Faul, F.; Erdfelder, E.; Lang, A.-G.; Buchner, A. G*Power 3: A flexible statistical power analysis program for the social, behavioral, and biomedical sciences. *Behav. Res. Methods* **2007**, *39*, 175–191. [CrossRef] [PubMed]
14. Alberti, K.G.; Zimmet, P.; Shaw, J. Metabolic syndrome—A new world-wide definition. A Consensus Statement from the International Diabetes Federation. *Diabet. Med.* **2006**, *23*, 469–480. [CrossRef] [PubMed]
15. Laughlin, M.H. Physical activity-induced remodeling of vasculature in skeletal muscle: Role in treatment of type 2 diabetes. *J. Appl. Physiol.* **2016**, *120*, 1–16. [CrossRef] [PubMed]
16. van Dijk, J.W.; Manders, R.J.; Tummers, K.; Bonomi, A.G.; Stehouwer, C.D.; Hartgens, F.; van Loon, L.J. Both resistance- and endurance type exercise reduce the prevalence of hyperglycaemia in individuals with impaired glucose tolerance and in insulin-treated and non-insulin-treated type 2 diabetic patients. *Diabetologia* **2012**, *55*, 1273–1282. [CrossRef]
17. Colberg, S.R.; Sigal, R.J.; Fernhall, B.; Regensteiner, J.G.; Blissmer, B.J.; Rubin, R.R.; Chasan-Taber, L.; Albright, A.L.; Braun, B.; American College of Sports Medicine, American Diabetes Association. Exercise and type 2 diabetes: The American College of Sports Medicine and the American Diabetes Association: Joint position statement. *Diabetes Care* **2010**, *33*, e147–e167. [CrossRef]
18. Colberg, S.R.; Sigal, R.J.; Yardley, J.E.; Riddell, M.C.; Dunstan, D.W.; Dempsey, P.C.; Horton, E.S.; Castorino, K.; Tate, D.F. Physical activity/exercise and diabetes: A position statement of the American Diabetes Association. *Diabetes Care* **2016**, *39*, 2065–2079. [CrossRef]
19. Mauvais-Jarvis, F. Gender differences in glucose homeostasis and diabetes. *Physiol. Behav.* **2018**, *187*, 20–23. [CrossRef]
20. van Genugten, R.E.; Utzschneider, K.M.; Tong, J.; Gerchman, F.; Zraika, S.; Udayasankar, J.; Boyko, E.J.; Fujimoto, W.Y.; Kahn, S.E. American Diabetes Association GENNID Study Group. Effects of sex and hormone replacement therapy use on the prevalence of isolated impaired fasting glucose and isolated impaired glucose tolerance in subjects with a family history of type 2 diabetes. *Diabetes* **2006**, *55*, 3529–3535. [CrossRef]
21. Nordestgaard, B.; Benn, M.; Schnohr, P.; Tybjaerg-Hansen, A. Nonfasting triglycerides and risk of myocardial infarction, ischemic heart disease, and death in men and women. *JAMA* **2007**, *289*, 299–308. [CrossRef]
22. Blum, S.; Aviram, M.; Ben-Amotz, A.; Levy, Y. Effect of a Mediterranean meal on post-prandial carotenoids, paraoxonase activity and C-reactive protein levels. *Ann. Nutr. Metab.* **2006**, *50*, 20–24. [CrossRef]
23. Jakulj, F.; Zernicke, K.; Bacon, S.L.; van Wielingen, L.E.; Key, B.L.; West, S.G.; Campbell, T.S. A high fat meal increases cardiovascular reactivity to psychological stress in healthy young adults. *J. Nutr.* **2007**, *137*, 935–939. [CrossRef] [PubMed]
24. Sattar, N.; Gaw, A.; Scherbakova, O. Metabolic syndrome with and without creactive protein as a predictor of coronary heart disease and diabetes in the West of Scotland Coronary Prevention Study. *Circulation* **2003**, *108*, 414–419. [CrossRef] [PubMed]

Article

The Assessment and Relationship Between Quality of Life and Physical Activity Levels in Greek Breast Cancer Female Patients under Chemotherapy

Maria Maridaki [1,*], Argyro Papadopetraki [2], Helen Karagianni [2], Michael Koutsilieris [2] and Anastassios Philippou [2]

[1] Faculty of Physical Education and Sport Science, National and Kapodistrian University of Athens, 17237 Dafne, Greece

[2] Medical School, National and Kapodistrian University of Athens, 11527 Athens, Greece; argpapa@med.uoa.gr (A.P.); elenikara13@gmail.com (H.K.); mkoutsil@med.uoa.gr (M.K.); tfilipou@med.uoa.gr (A.P.)

* Correspondence: mmarida@phed.uoa.gr; Tel.: +30-210-727

Received: 30 January 2020; Accepted: 9 March 2020; Published: 11 March 2020

Abstract: A growing body of evidence suggests that physical activity (PA) can be a complementary intervention during breast cancer (BCa) treatment, contributing to the alleviation of the chemotherapy-related side-effects. The purpose of this study was to assess physical activity (PA) levels and quality of life (QoL) parameters of BCa patients undergoing chemotherapy and compare them with healthy controls. A total of 94 BCa female patients and 65 healthy women were recruited and self-reported QoL and PA levels. The results reveal that women suffering from BCa spent only 134 ± 469 metabolic equivalents (MET)/week in vigorous PAs compared with the healthy females who spent 985±1508 MET/week. Also, BCa patients were spending 4.62±2.58 h/day sitting, contrary to the 2.34±1.05 h/day of the controls. QoL was scored as 63.43±20.63 and 70.14±19.49 while physical functioning (PF) as 71.48±23.35 and 84.46±15.48 by BCa patients and healthy participants, respectively. Negative correlations were found between QoL and fatigue, PF and pain, and fatigue and dyspnea, while a positive correlation was found between QoL and PF. This study indicated that the BCa group accumulated many hours seated and refrained from vigorous Pas, preferring PAs of moderate intensity. Additionally, BCa patients' levels of functioning and QoL were moderate to high; however, they were compromised by pain, dyspnea and fatigue.

Keywords: breast cancer; chemotherapy; physical activity; quality of life; exercise; QoL; treatment

---

1. Introduction

According to the World Health Organization (WHO), cancer is a leading cause of mortality worldwide, while approximately one out of six deaths is due to cancer. In both sexes, lung cancer is the most commonly diagnosed malignancy and the most frequent cancer leading to death. On the other hand, among females, breast cancer (BCa) constitutes the most commonly diagnosed cancer, as well as the first in mortality rate [1]. Epidemiological studies revealed that in spite of the fact that BCa accounts for about 30% of all cancer diagnoses in women [2], the overall 5-year survival rate is over 90% for survivors diagnosed with BCa stage I or II [3].

The increased survival rates due to advancements in cancer detection and medical care indicate that cancer should be handled as a chronic disease that requires long term management to maintain patients' quality of life [4]. It is well established that standard medical care for BCa, including surgery, chemotherapy, radiotherapy and hormonal therapy, is associated with adverse effects on cardiorespiratory, musculoskeletal, nervous and endocrine physiological systems [5–7]. In particular, cardiotoxicity, cancer-related fatigue, muscle atrophy, cachexia, peripheral neuropathy, immune system dysfunction and altered body composition are some of the reported complications that result in a diminished quality of life (QoL) of patients, while interfering with their ability to carry out regular daily living activities [8–11]. Moreover, general pain and fatigue belong to the most frequently experienced symptoms that cancer patients undergoing treatment exhibit and these symptoms are related to the severity of the disease [12].

However, an increasing body of evidence suggests that prescribed exercise during and after cancer treatment may attenuate many of these adverse effects and mitigate several symptoms, constituting a safe complementary therapeutic intervention for cancer patients [12]. In addition to the studies that suggest the preventive role of physical activity against BCa risk [13,14], there is also evidence supporting that regular exercise also reduces the risk of disease recurrence for several types of solid tumors including BCa. These inhibitory effects of regular exercise are probably mediated by different mechanisms that alter the tumor microenvironment [15,16].

The American College of Sports Medicine (ACSM) and the American Cancer Society (ACS) recommend that BCa patients should avoid remaining inactive and aim to return to their normal daily routine as soon as possible after diagnosis and during the treatment of the disease. For instance, BCa patients should be encouraged to accumulate at least 150 or 75 min of moderate or vigorous aerobic exercise per week, respectively, and include resistant training exercises two to three times per week [17]. The compliance to these guidelines is really important for the individuals subjected to cancer treatment, because, as in a chronic disease, so in cancer, there is a dose–response relationship between physical activity (PA) levels and health benefits gained [18].

Despite the abovementioned recommendations, current research evidence suggests that the majority of people living with cancer do not participate in PAs and they adopt sedentary behavior [19,20]. The purpose of the present study was to assess QoL and PA levels of BCa female patients living in Greece and undergoing chemotherapy, and to compare them with healthy age- and sex-matched controls.

2. Materials and Methods

2.1. Ethical Approval

All volunteers provided written informed consent to participate in this cross-sectional observational study, which was approved by the seven-member Ethics Committee of the Medical School of the National and Kapodistrian University of Athens. All data were collected and handled according to privacy law regulations.

2.2. Subjects

A total of 159 females, aged from 42 to 71 years, voluntarily participated in the study. From them, 94 women (age: 57.25 ± 13.59 years) were newly diagnosed with breast cancer for first time, stage I-III, and had already started to receive first-line chemotherapy, while 65 healthy women (age: 49.60 ± 7.80 years) served as a control group. The patients were recruited in close collaboration with the attendant physicians from three different Greek hospitals, from the October of 2017 to the October of 2018, and filled the questionnaires during their first regimen of chemotherapy while no exclusion criteria were set according to the type of surgery that had preceded. The women who comprised the control group were recruited in the same chronological period and they should have never been diagnosed with cancer. Moreover, all participants should speak and read Greek.

2.3. Data Collection

All participants filled in the structured questionnaires, while their body height and body mass were measured in order for their body mass index (BMI) to be calculated. Participants were instructed to answer all the questions as carefully and honestly as possible, while an investigator was available for providing clarifications for any possible questions raised regarding the way that the questionnaires should be filled in.

2.3.1. Somatometric Characteristics

Body height was measured with the subject standing in bare feet with her back towards a height measuring rod and body mass was measured with an electronic precision balance with two decimals. Body mass index (BMI) was then calculated according to the following formula: BMI = body mass (kg) / body height ^2 (m^2). Individuals were considered to be of normal body weight if their BMI was between 20 and 24.9, while they were considered as underweight if their BMI was lower than 20. If BMI was in the range between 25 and 29.9, or higher than 30, the individual was considered as overweight or obese, respectively [21].

2.3.2. Quality of Life

Quality of life was self-estimated by the BCa patients and the healthy controls, using the EORTQ-QLQ-C30 or the SF-36 Health Survey Version 3.0 questionnaire, respectively [22–24]. Specifically, EORTQ-QLQ-C30 is a cancer-specific questionnaire that incorporates global health status/QoL scale, common symptom scales and physical, emotional, cognitive, role and social functioning scales. For example, some of the items the questionnaire focuses on are pain, fatigue, sleep, concentration, appetite etc. In this particular questionnaire, the score in each scale ranges from 0 to 100. The higher the score on the functional scales or the global health status is, the greater the level of functioning and QoL. Reversely, a high score in the symptom scale reflects a high level of symptomatology.

Similarly, SF-36 is a 36-item questionnaire that covers eight health domains: physical functioning, pain, fatigue, role limitations due to physical health problems, role limitations due to emotional problems, emotional well-being, social functioning and general health perceptions. Each item of this questionnaire is also scored on a 0 to 100 scale and in all scales a higher score defines a more favorable health status. For instance, a higher score in the fatigue scale actually represents less fatigue. The two questionnaires, SF-36 and EORTQ-QLQ-C30, have the same way of scoring and interpreting the results in the scales general QoL, Physical Functioning, Emotional Functioning, Social Functioning and Role Functioning. Thus, the comparisons of QoL were based on the similarity in scales (0–100, with a higher score indicating better health) and not on actual survey questions or summary calculations.

2.3.3. Exercise Behavior

Current PA levels were self-reported by the participants using the short version of the International Physical Activity Questionnaire (IPAQ). IPAQ assesses the duration and the intensity of PAs as well as the time spent sitting in daily lives, while it is considered to estimate the total weekly energy expenditure in MET-min per week. Activities that require up to 3 METs have been defined as light-intensity PAs, activities that range from 3 to 6 METs have been categorized as moderate-intensity PAs, whereas those that require more than 6 METs were defined as vigorous-intensity PAs [25].

2.4. Statistical Analysis

Statistical analysis was conducted using Graphpad Prism Version 5.03 (GraphPad Software, Inc., San Diego, CA, USA). For all quantitative variables, descriptive analysis was employed by mean and standard deviation (MEAN ± SD), while evaluation of the potential differences between the two independent groups (i.e., BCa vs Control group) was performed with a two-tailed, unpaired Student T-test. Pearson parametric correlation coefficient was utilized to determine any potential associations between the continuous variables—physical activity and QoL. The level of statistical significance was set at $P < 0.05$.

3. Results

3.1. Somatometric Characteristics

The somatometric characteristics of the participants in each group (i.e., BCa patients and healthy controls) are shown in Table 1. Height was 1.61 ± 0.05 m and 1.65 ± 0.04 m in the BCa and control group, respectively, while body mass was 69.49 ± 12.67 kg in BCa patients and 69.04 ± 5.25 kg in healthy controls. BMI was used for the classification of participants as underweight, normoweight, overweight or obese. These results reveal that BCa patients' BMI was 26.63 ± 5.27 kg/m^2, categorizing them as overweight, by contrast with the healthy females in the control group whose BMI was marginally normal (25.30 ± 3.95 kg/m^2).

Table 1. Somatometric characteristics of breast cancer patients and healthy participants (control group).

Participants' Characteristics	Breast Cancer Group (n=94)	Control Group (n=65)
Age (yrs)	57.25 ± 13.59	49.60 ± 7.80
Body Mass (kg)	69.49 ± 12.67	69.04 ± 5.25
Body Height (m)	1.61 ± 0.05	1.65 ± 0.04
Body Mass Index (kg/m^2)	26.63 ± 5.27	25.30 ± 3.95

Data are presented as mean ± SD. No statistically significant differences were found between groups ($p>0.05$).

3.2. Quality of Life

3.2.1. Control Group

Healthy females who served as the control group self-evaluated their QoL using the SF-36 Health Survey Version 3.0. Regarding their general QoL, the participants scored 70.14 ± 19.49, while for their physical, emotional, social and role functioning their score was 84.46 ± 15.48, 59.33 ± 17.83, 61.79 ± 27.05 and 79.17 ± 29.76, respectively (Figure 1). Moreover, in the symptom scales, pain was scored with 70.42 ± 22.93 and fatigue with 58.06 ± 12.23. Positive correlations were revealed between physical functioning and pain ($r = 0.4432$, $p = 0.007$), fatigue ($r = 0.4847$, $p = 0.003$), emotional functioning ($r = 0.4133$, $p = 0.012$) and role functioning ($r = 0.3869$, $p = 0.020$). Positive correlations were also found between QoL and the scales of physical functioning ($r = 0.4072$, $p = 0.014$) and fatigue ($r = 0.6653$, $p = 0.00001$), (Figure 2).

3.2.2. Breast Cancer Group

Similarly to the control group, the women of the BCa group self-estimated their QoL using the EORTC-QLQ-C30 Questionnaire. Women in the BCa group scored their physical functioning significantly lower compared with the healthy controls (71.48 ± 23.35 vs 84.46 ± 15.48; $p<0.01$). However, their overall QoL, as well as their emotional, social and role functioning score, was 63.43 ± 20.63, 67.13 ± 27.02, 68.52 ± 31.31 and 68.98 ± 26.77, respectively, revealing no significant differences with the control group ($p > 0.05$) (Figure 1). It is noted that comparisons between the BCa and control group were performed only between the above-mentioned scales, since the rest of them in each questionnaire have a different way of scoring.

As far the symptomatology is concerned, fatigue was scored at 42.28 ± 20.54, dyspnea at 25.93 ± 28.85 and pain at 19.44 ± 24.40. A negative correlation was found between QoL and fatigue ($r = -0.7410$, $p = 0.00001$), as well as between physical functioning and pain ($r = -0.6149$, $p = 0.0001$), fatigue ($r = -0.6661$, $p = 0.0001$) and dyspnea ($r = -0.3320$, $p = 0.0493$), (Figure 2). In contrast, a positive correlation was revealed between physical functioning and QoL ($r = 0.4914$, $p = 0.0024$), social functioning ($r = 0.5954$, $p = 0.0001$) and emotional functioning ($r = 0.3663$, $p = 0.0263$) (Figure 2).

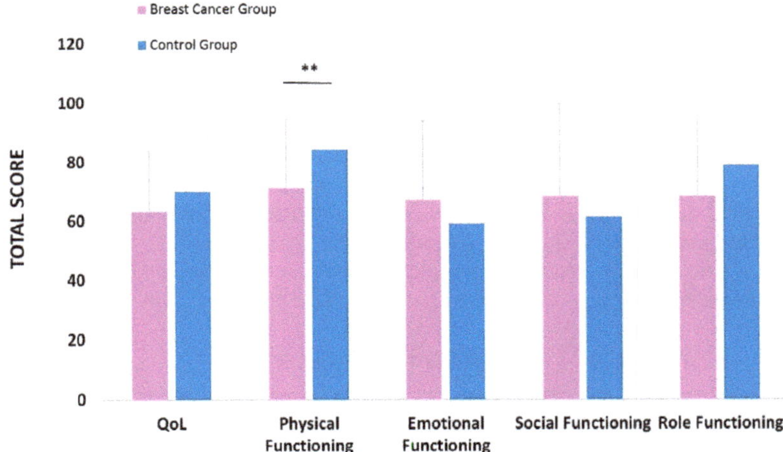

Figure 1. Self-estimation of the overall quality of life (QoL) and its functional parameters in women undergoing chemotherapy for breast cancer compared with healthy controls. Data are presented as mean ± SD. **: Significantly different at $p < 0.01$.

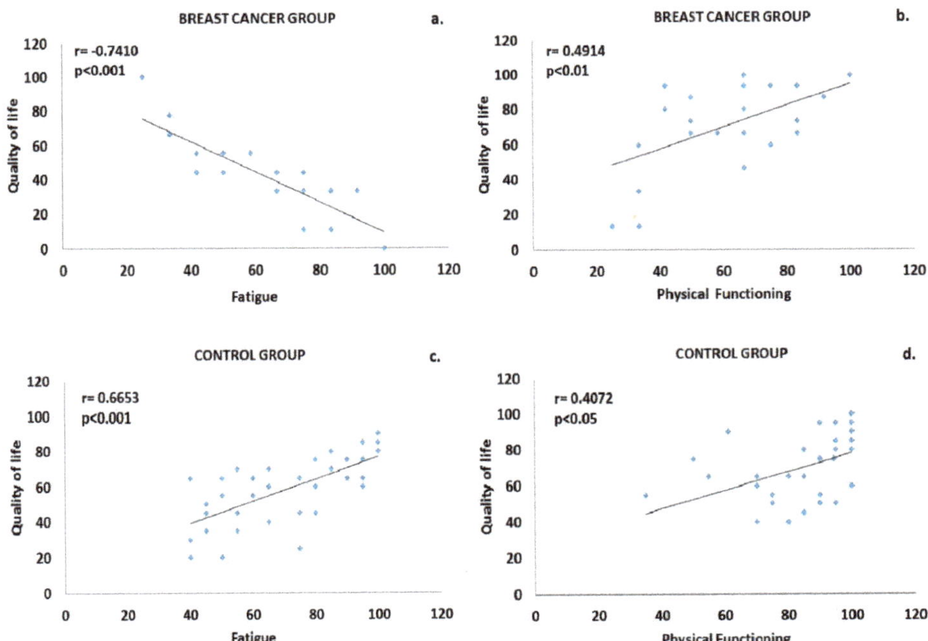

Figure 2. Correlational analyses revealed significant associations, among others (see text for details), between fatigue and quality of life (**a**,**c**), as well as between physical functioning and quality of life (**b**,**d**), both in the breast cancer and the control group.

3.3. Exercise Behavior

Exercise behavior was self-reported by all participants using the International Physical Activity Questionnaire (IPAQ) (Figure 3). Specifically, BCa patients exhibited a total energy expenditure of 2267 ± 1965 MET-min/week, while healthy controls spent 2630 ± 2840 MET-min/week, showing no significant differences between groups ($p > 0.05$). In particular, no significant differences ($p > 0.05$) were found between the two groups in the time spent walking (BCa group: 782 ± 1,153 MET-min/week vs Control group: 721 ± 950 MET-min/week). A similar ($p > 0.05$) energy expenditure was also spent in moderate PAs by both BCa and control group, i.e., 1460 ± 1549 vs 1089 ± 1724 MET-min/week, respectively. Interestingly, on the other hand, BCa patients were found to participate in vigorous PAs disproportionally less than the control group, expending only 134 ± 469 MET-min/week, as opposed to the control group that spent 985 ± 1,508 MET-min/week in high-intensity activities ($p < 0.001$). It is noted that moderate PAs require intermediate physical effort and make breathing somewhat harder than normal, while vigorous PAs need excess physical effort, increasing breath rate.

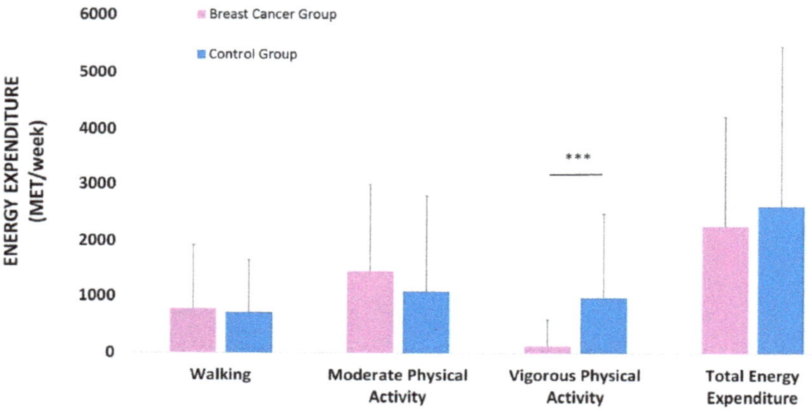

Figure 3. Self-reported physical activity levels (International Physical Activity Questionnaire (IPAQ)) in women undergoing chemotherapy for breast cancer compared with healthy controls, expressed in MET-min per week. Data are presented as mean ± SD.***: Significantly different at $p<0.001$.

Furthermore, BCa patients were found to spend more time sitting during the day (4.20 ± 2.76 h/day) in comparison with the control group (3.16 ± 1.25 h/day), ($p < 0.05$). Again, it is noted that sedentary time includes time spent sitting or lying down during work and leisure, or at home and excludes sleeping hours.

3.4. Associations between Exercise Behavior and Quality of Life

In the BCa group, a positive correlation was demonstrated between physical functioning and total energy expenditure ($r = 0.4069$, $p = 0.0316$) (Figure 4a), as well as between QoL and participation in vigorous PAs ($r = 0.3985$, $p = 0.0357$). Similarly, a positive correlation was also found in the control group between the engagement in vigorous PAs and QoL ($r = 0.4993$, $p = 0.0094$) (Figure 4b), as well as between vigorous PAs and physical functioning ($r = 0.5149$, $p = 0.0071$).

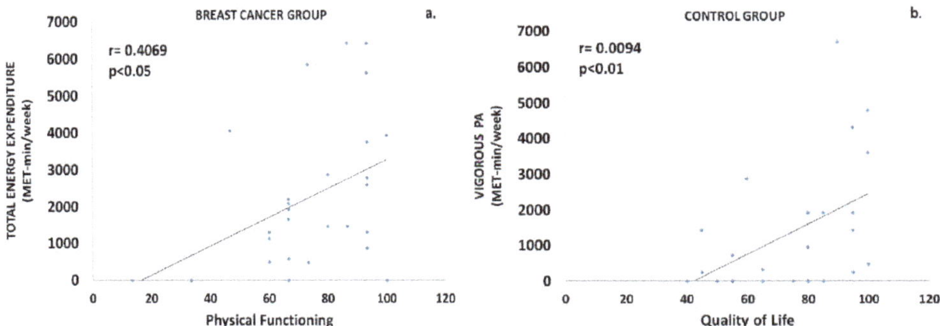

Figure 4. Correlational analyses showed significant associations, among others (see text for details), between (**a**) total energy expenditure and physical functioning in the breast cancer group as well as between (**b**) the engagement in vigorous PAs and the quality of life, in the control group.

4. Discussion

The aim of the present study was to identify the levels of PA and the perceived QoL, investigating their potential interactions, in females undergoing chemotherapy due to BCa diagnosis, and to compare them with healthy females of the same age.

Our main findings demonstrate that women suffering from BCa and undergoing chemotherapy were willing to exercise and they participated in regular PAs, exhibiting weekly energy expenditure levels similar to those of sex- and aged-matched healthy individuals. However, they preferred to exercise in low or moderate intensities, showing significantly lower levels of MET-min per week expended in high intensity PAs compared with the healthy controls. These findings are in agreement with previous studies implying that cancer patients demonstrate lower levels of vigorous-intensity PAs post than before diagnosis [26,27]. Even though it has been established that high-intensity activities can safely be performed by cancer patients, offering different health benefits than those derived from the conventional exercise programs, cancer patients appear to hesitate to participate in vigorous PAs [28]. On the other hand, cancer-related fatigue and general pain probably exacerbate the overall burden of the disease and the therapeutic interventions, making participation in more intense physical activities difficult, especially for those patients with more advanced stages of the disease.

Moreover, our study showed that although the BCa patients were exercising in general, they accumulated many hours per day sitting down, not only at work but also at home, since many patients often interrupted their work during chemotherapy sessions, thus spending more hours per day seated at home, which may result in their overweight phenotype. These findings strengthen the evidence from previous studies which supported the hypothesis that an increased BMI is associated with a sedentary lifestyle after cancer diagnosis [29]. Since an increased body weight has been associated with a higher risk of disease recurrence and reduced survival, all cancer patients should not only avoid remaining physically inactive but also they need to follow the specific exercise recommendations, so as to optimize their health exercise benefits [30].

Regarding the QoL, our study showed that BCa patients exhibited moderate levels of perceived QoL, similarly to the control group. More specifically, a strong negative correlation was found between QoL and fatigue as expected, highlighting the fact that cancer-related fatigue remains a huge barrier to patients' daily life [31,32]. In addition, negative correlations were also found between physical functioning and the side effects of the disease, such as pain, fatigue and dyspnea, indicating that these symptoms compromise patients' functional capacity and QoL [33]. Similar associations between the symptomatology and functional scales were also observed in the control group, indicating that the above-mentioned clinical symptoms influence the individuals' daily life independently of the disease.

Moreover, with regard to the relationship between exercise behavior and QoL, a positive correlation was found between participation in vigorous PAs and QoL, as well as between total energy expenditure and physical functioning in women with BCa. These findings corroborate a large body of evidence supporting the hypothesis that a greater energy expenditure during the week leads to a better functional ability, while participation in more intense activities implies a better self-evaluated QoL [34,35]. Moreover, the strong positive correlations found between physical, social and emotional functioning further support previous findings that mental health symptoms and isolation are followed by a poor functional ability in cancer patients [36,37].

Putting all the above findings together, it appears that new approaches are urgently needed to improve tolerance and reduce the adverse effects of chemotherapy in cancer patients [38]. Physical activity interventions should be incorporated in cancer non-pharmaceutical treatments during chemotherapy, since the worst side effects of cancer therapy are experienced during this period, while exercise can mitigate unfavorable changes in various physiological systems and their consequent symptoms [39–41]. Clinical physicians are proposed to assess, advise and refer cancer patients to exercise [19,42,43].

5. Conclusions and Future Perspectives

The outcomes of the present study unveil a close relationship between exercise behavior and QoL in breast cancer patients; however, there remain challenging issues to be further addressed. Future research lines of investigation should focus on the dose-dependent effects of physical activity and on revealing the optimum dose as well as the potential maximum and minimum thresholds of the cancer patients for benefit from physical activity. Furthermore, it remains a challenge to elucidate whether cancer type, timing of physical activity and its specific components influence the effectiveness of exercise and its interactions with cancer outcomes. For instance, in order for vigorous physical activities to be realistically adopted and sustained by those patients during their treatment, a mode of short-duration high-intensity physical exercise with adequate breaks might be a more applicable suggestion for them, so as to take advantage of the time-effective, beneficial effects of vigorous activities on their quality of life and physical functioning. Since physical activity is an important factor for cancer prevention and treatment, policy makers, public health professionals, health care providers, and exercise scientists should all communicate and promote the benefits of physical activity for both cancer prevention and control, and work together with other stakeholders to improve the health and quality of life of cancer patients.

Author Contributions: Conceptualization: M.M., A.P. (Argyro Papadopetraki), H.K., A.P. (Anastassios Philippou); methodology: M.M. and A.P. (Anastassios Philippou); investigation: H.K., A.P. (Argyro Papadopetraki); data curation: A.P. (Argyro Papadopetraki), H.K; writing—original draft preparation: A.P. (Argyro Papadopetraki); writing—review and editing: M.M., A.P. (Anastassios Philippou), M.K.; supervision: M.M., A.P. (Anastassios Philippou). All authors have read and agreed to the published version of the manuscript.

Funding: This research received no external funding.

Acknowledgments: The authors are grateful to the study participants for their invaluable contribution to this study.

Conflicts of Interest: The authors declare no conflict of interest.

References

1. Bray, F.; Ferlay, J.; Soerjomataram, I.; Siegel, R.L.; Torre, L.A.; Jemal, A. Global cancer statistics 2018: GLOBOCAN estimates of incidence and mortality worldwide for 36 cancers in 185 countries. *CA Cancer J. Clin.* **2018**, *68*, 394–424. [CrossRef] [PubMed]
2. Siegel, R.L.; Miller, K.D.; Jemal, A. Cancer statistics, 2019. *CA Cancer J. Clin.* **2019**, *69*, 7–34. [CrossRef] [PubMed]
3. DeSantis, C.E.; Ma, J.; Gaudet, M.M.; Newman, L.A.; Miller, K.D.; Sauer, A.G.; Jemal, A.; Siegel, R.L. Breast cancer statistics, 2019. *CA Cancer J. Clin.* **2019**, *69*, 438–451. [CrossRef] [PubMed]

4. Devlin, E.J.; Denson, L.A.; Whitford, H.S. Cancer Treatment Side Effects: A Meta-analysis of the Relationship between Response Expectancies and Experience. *J. Pain Symptom Manag.* **2017**, *54*, 245–258.e2. [CrossRef]
5. Stefani, L.; Giorgio, G.; Klika, R. Clinical Implementation of Exercise Guidelines for Cancer Patients: Adaptation of ACSM's Guidelines to the Italian Model. *J. Funct. Morphol. Kinesiol.* **2017**, *2*, 4. [CrossRef]
6. Neil-Sztramko, S.E.; Winters-Stone, K.M.; Bland, K.A.; Campbell, K.L. Updated systematic review of exercise studies in breast cancer survivors: Attention to the principles of exercise training. *Br. J. Sports Med.* **2019**, *53*, 504–512. [CrossRef]
7. Adraskela, K.; Veisaki, E.; Koutsilieris, M.; Philippou, A. Physical Exercise Positively Influences Breast Cancer Evolution. *Clin. Breast Cancer* **2017**, *17*, 408–417. [CrossRef]
8. Bao, T.; Basal, C.; Seluzicki, C.; Li, S.Q.; Seidman, A.D.; Mao, J.J. Long-term chemotherapy-induced peripheral neuropathy among breast cancer survivors: Prevalence, risk factors, and fall risk. *Breast Cancer Res. Treat.* **2016**, *159*, 327–333. [CrossRef]
9. Howden, E.J.; Bigaran, A.; Beaudry, R.; Fraser, S.; Selig, S.; Foulkes, S.; Antill, Y.; Nightingale, S.; Loi, S.; Haykowsky, M.J.; et al. Exercise as a diagnostic and therapeutic tool for the prevention of cardiovascular dysfunction in breast cancer patients. *Eur. J. Prev. Cardiol.* **2019**, *26*, 305–315. [CrossRef]
10. Schirrmacher, V. From chemotherapy to biological therapy: A review of novel concepts to reduce the side effects of systemic cancer treatment (Review). *Int. J. Oncol.* **2019**, *54*, 407–419.
11. Argiles, J.M.; Stemmler, B.; López-Soriano, F.J.; Busquets, S. Inter-tissue communication in cancer cachexia. *Nat. Rev. Endocrinol.* **2018**, *15*, 9–20. [CrossRef] [PubMed]
12. Wonders, K.Y.; Wise, R.; Ondreka, D.; Seitz, T. Supervised, Individualized Exercise Mitigates Symptom Severity during Cancer Treatment. *J. Adenocarcinoma Osteosarcoma* **2018**, *3*, 1–5.
13. Guo, W.; Fensom, G.K.; Reeves, G.K.; Key, T.J. Physical activity and breast cancer risk: Results from the UK Biobank prospective cohort. *Br. J. Cancer* **2020**, *122*, 726–732. [CrossRef] [PubMed]
14. Kehm, R.D.; Genkinger, J.M.; MacInnis, R.J.; John, E.M.; Phillips, K.A.; Dite, G.S.; Milne, R.L.; Zeinomar, N.; Liao, Y.; Knight, J.A.; et al. Recreational Physical Activity Is Associated with Reduced Breast Cancer Risk in Adult Women at High Risk for Breast Cancer: A Cohort Study of Women Selected for Familial and Genetic Risk. *Cancer Res.* **2020**, *80*, 116–125. [CrossRef] [PubMed]
15. Koelwyn, G.J.; Quail, D.F.; Zhang, X.; White, R.M.; Jones, L.W. Exercise-dependent regulation of the tumour microenvironment. *Nat. Rev. Cancer* **2017**, *17*, 620–632. [CrossRef] [PubMed]
16. Hojman, P.; Gehl, J.; Christensen, J.F.; Pedersen, B.K. Molecular Mechanisms Linking Exercise to Cancer Prevention and Treatment. *Cell Metab.* **2018**, *27*, 10–21. [CrossRef]
17. Runowicz, C.D.; Leach, C.R.; Henry, N.L.; Henry, K.S.; Mackey, H.T.; Cowens-Alvarado, R.L.; Cannady, R.S.; Pratt-Chapman, M.L.; Edge, S.B.; Jacobs, L.A.; et al. American Cancer Society/American Society of Clinical Oncology Breast Cancer Survivorship Care Guideline. *CA Cancer J. Clin.* **2016**, *66*, 43–73. [CrossRef]
18. Durstine, J.L.; Gordon, B.; Wang, Z.; Luo, X. Chronic disease and the link to physical activity. *J. Sport Health Sci.* **2013**, *2*, 3–11. [CrossRef]
19. Schmitz, K.H.; Campbell, A.M.; Stuiver, M.M.; Pinto, B.M.; Schwartz, A.L.; Morris, G.S.; Ligibel, J.A.; Cheville, A.; Galvão, D.A.; Alfano, C.M.; et al. Exercise is medicine in oncology: Engaging clinicians to help patients move through cancer. *CA Cancer J. Clin.* **2019**, *69*, 468–484. [CrossRef]
20. Ramirez-Parada, K.; Courneya, K.S.; Muñiz, S.; Sánchez, C.; Fernández-Verdejo, R. Physical activity levels and preferences of patients with breast cancer receiving chemotherapy in Chile. *Support Care Cancer* **2019**, *27*, 2941–2947. [CrossRef]
21. Nuttall, F.Q. Body Mass Index: Obesity, BMI, and Health: A Critical Review. *Nutr. Today* **2015**, *50*, 117–128. [CrossRef] [PubMed]
22. Hong, F.; Ye, W.; Kuo, C.H.; Zhang, Y.; Qian, Y.; Korivi, M. Exercise Intervention Improves Clinical Outcomes, but the "Time of Session" is Crucial for Better Quality of Life in Breast Cancer Survivors: A Systematic Review and Meta-Analysis. *Cancers* **2019**, *11*, 706. [CrossRef] [PubMed]
23. Young-Afat, D.A.; Van Gils, C.H.; Van Den Bongard, H.J.G.D.; Verkooijen, H.M.; UMBRELLA Study Group. The Utrecht cohort for Multiple BREast cancer intervention studies and Long-term evaLuAtion (UMBRELLA): Objectives, design, and baseline results. *Breast Cancer Res. Treat.* **2017**, *164*, 445–450. [CrossRef] [PubMed]
24. Verket, N.J.; Uhlig, T.; Sandvik, L.; Andersen, M.H.; Tanbo, T.G.; Qvigstad, E. Health-related quality of life in women with endometriosis, compared with the general population and women with rheumatoid arthritis. *Acta Obstet. Gynecol. Scand.* **2018**, *97*, 1339–1348. [CrossRef]

25. Irwin, M.L.; Aiello, E.J.; McTiernan, A.; Baumgartner, R.N.; Baumgartner, K.B.; Bernstein, L.; Gilliland, F.D.; Ballard-Barbash, R. Pre-diagnosis physical activity and mammographic density in breast cancer survivors. *Breast Cancer Res. Treat.* **2006**, *95*, 171–178. [CrossRef]
26. Fassier, P.; Zelek, L.; Partula, V.; Srour, B.; Bachmann, P.; Touillaud, M.; Druesne-Pecollo, N.; Galan, P.; Cohen, P.; Hoarau, H.; et al. Variations of physical activity and sedentary behavior between before and after cancer diagnosis: Results from the prospective population-based NutriNet-Sante cohort. *Medicine (Baltim.)* **2016**, *95*, e4629. [CrossRef]
27. Littman, A.J.; Tang, M.T.; Rossing, M.A. Longitudinal study of recreational physical activity in breast cancer survivors. *J. Cancer Surviv.* **2010**, *4*, 119–127. [CrossRef]
28. Papadopoulos, E.; Mina, D.S. Can we HIIT cancer if we attack inflammation? *Cancer Causes Control.* **2018**, *29*, 7–11. [CrossRef]
29. Harrison, S.; Hayes, S.C.; Newman, B. Level of physical activity and characteristics associated with change following breast cancer diagnosis and treatment. *Psychooncology* **2009**, *18*, 387–394. [CrossRef]
30. Chan, D.S.; Norat, T. Obesity and breast cancer: Not only a risk factor of the disease. *Curr. Treat. Options Oncol.* **2015**, *16*, 22. [CrossRef]
31. Meneses-Echavez, J.F.; Gonzalez-Jimenez, E.; Ramirez-Velez, R. Effects of supervised exercise on cancer-related fatigue in breast cancer survivors: A systematic review and meta-analysis. *BMC Cancer* **2015**, *15*, 77. [CrossRef] [PubMed]
32. Bower, J.E. Cancer-related fatigue—Mechanisms, risk factors, and treatments. *Nat. Rev. Clin. Oncol.* **2014**, *11*, 597–609. [CrossRef] [PubMed]
33. Pearce, A.; Haas, M.; Viney, R.; Pearson, S.A.; Haywood, P.; Brown, C.; Ward, R. Incidence and severity of self-reported chemotherapy side effects in routine care: A prospective cohort study. *PLoS ONE* **2017**, *12*, e0184360. [CrossRef] [PubMed]
34. Schwartz, A.L.; de Heer, H.D.; Bea, J.W. Initiating Exercise Interventions to Promote Wellness in Cancer Patients and Survivors. *Oncology (Williston Park)* **2017**, *31*, 711–717. [PubMed]
35. Kirkham, A.A.; Bland, K.A.; Sayyari, S.; Campbell, K.L.; Davis, M.K. Clinically Relevant Physical Benefits of Exercise Interventions in Breast Cancer Survivors. *Curr. Oncol. Rep.* **2016**, *18*, 12. [CrossRef] [PubMed]
36. Heidari, M.; Ghodusi, M. The Relationship between Body Esteem and Hope and Mental Health in Breast Cancer Patients after Mastectomy. *Indian J. Palliat. Care* **2015**, *21*, 198–202.
37. Cvetkovic, J.; Nenadovic, M. Depression in breast cancer patients. *Psychiatry Res.* **2016**, *240*, 343–347. [CrossRef]
38. Nurgali, K.; Jagoe, R.T.; Abalo, R. Editorial: Adverse Effects of Cancer Chemotherapy: Anything New to Improve Tolerance and Reduce Sequelae? *Front. Pharmacol.* **2018**, *9*, 245. [CrossRef]
39. Ruiz-Casado, A.; Martín-Ruiz, A.; Pérez, L.M.; Provencio, M.; Fiuza-Luces, C.; Lucia, A. Exercise and the Hallmarks of Cancer. *Trends Cancer* **2017**, *3*, 423–441. [CrossRef]
40. Buffart, L.M.; Kalter, J.; Sweegers, M.G.; Courneya, K.S.; Newton, R.U.; Aaronson, N.K.; Jacobsen, P.B.; May, A.M.; Galvão, D.A.; Chinapaw, M.J.; et al. Effects and moderators of exercise on quality of life and physical function in patients with cancer: An individual patient data meta-analysis of 34 RCTs. *Cancer Treat. Rev.* **2017**, *52*, 91–104. [CrossRef]
41. Sweegers, M.G.; Altenburg, T.; Chinapaw, M.J.; Kalter, J.; Verdonck-de Leeuw, I.M.; Courneya, K.S.; Newton, R.U.; Aaronson, N.K.; Jacobsen, P.; Brug, J.; et al. Which exercise prescriptions improve quality of life and physical function in patients with cancer during and following treatment? A systematic review and meta-analysis of randomised controlled trials. *Br. J. Sports Med.* **2018**, *52*, 505–513. [CrossRef] [PubMed]
42. Short, C.E.; Rebar, A.; James, E.L.; Duncan, M.J.; Courneya, K.S.; Plotnikoff, R.C.; Crutzen, R.; Vandelanotte, C. How do different delivery schedules of tailored web-based physical activity advice for breast cancer survivors influence intervention use and efficacy? *J. Cancer Surviv.* **2017**, *11*, 80–91. [CrossRef] [PubMed]
43. Ferrer, R.A.; Huedo-Medina, T.B.; Johnson, B.T.; Ryan, S.; Pescatello, L.S. Exercise interventions for cancer survivors: A meta-analysis of quality of life outcomes. *Ann. Behav. Med.* **2011**, *41*, 32–47. [CrossRef] [PubMed]

MDPI
St. Alban-Anlage 66
4052 Basel
Switzerland
Tel. +41 61 683 77 34
Fax +41 61 302 89 18
www.mdpi.com

Sports Editorial Office
E-mail: sports@mdpi.com
www.mdpi.com/journal/sports

www.ingramcontent.com/pod-product-compliance
Lightning Source LLC
LaVergne TN
LVHW070626170726
843515LV00004B/189

* 9 7 8 3 0 3 6 5 0 4 2 8 5 *